CULTURE, HERITAGE AND REPRESENTATION

Heritage, Culture and Identity

Series Editor: Brian Graham,
School of Environmental Sciences, University of Ulster, UK

Other titles in this series

Sport, Leisure and Culture in the Postmodern City
Edited by Peter Bramham and Stephen Wagg
ISBN 978 0 7546 7274 6

Valuing Historic Environments
Edited by Lisanne Gibson and John Pendlebury
ISBN 978 0 7546 7424 5

Southeast Asian Culture and Heritage in a Globalising World
Diverging Identities in a Dynamic Region
Edited by Rahil Ismail, Brian Shaw and Ooi Giok Ling
ISBN 978 0 7546 7261 6

Jewish Topographies
Visions of Space, Traditions of Place
Edited by Julia Brauch, Anna Lipphardt and Alexandra Nocke
ISBN 978 0 7546 7118 3

Geography and Genealogy
Locating Personal Pasts
Edited by Dallen J. Timothy and Jeanne Kay Guelke
ISBN 978 0 7546 7012 4

Living Ruins, Value Conflicts
Argyro Loukaki
ISBN 978 0 7546 7228 9

Geographies of Australian Heritages
Loving a Sunburnt Country?
Edited by Roy Jones and Brian J. Shaw
ISBN 978 0 7546 4858 1

Culture, Heritage and Representation

Perspectives on Visuality and the Past

Edited by

EMMA WATERTON
Keele University, UK

and

STEVE WATSON
York St John University, UK

ASHGATE

Reprinted 2011

Published by
Ashgate Publishing Limited
Wey Court East
Union Road
Farnham
Surrey, GU9 7PT
England

Ashgate Publishing Company
Suite 420
101 Cherry Street
Burlington
VT 05401-4405
USA

www.ashgate.com

British Library Cataloguing in Publication Data
Culture, heritage and representation : perspectives on
visuality and the past. -- (Heritage, culture and identity)
1. Memorialization. 2. Heritage tourism. 3. Visual
communication. 4. Image (Philosophy)
I. Series II. Waterton, Emma. III. Watson, Steve.
306.4'819-dc22

Library of Congress Cataloging-in-Publication Data
Culture, heritage and representation : perspectives on visuality and the past / [edited] by Emma Waterton and Steve Watson.
p. cm. -- (Heritage, culture and identity)
Includes bibliographical references and index.
ISBN 978-0-7546-7598-3 (hbk) 1. Heritage tourism--Case studies. 2. Visual anthropology--Case studies. I. Waterton, Emma. II. Watson, Steve.
G156.5.H47C8598 2010
338.4'791--dc22

2009046059

ISBN 9780754675983 (hbk)

Printed and bound in Great Britain by
TJI Digital, Padstow, Cornwall.

Contents

List of Figures

List of Tables

Notes on Contributors

Dr Tim Copeland is Head of the International Centre for Education at the University of Gloucestershire. He has been a head teacher and archaeologist (teaching at both Universities of Gloucestershire and Bristol), and is interested in visitors', especially children's, experiences of historic sites. Much of his research has been on constructivist approaches to evidence from the past as well as Iron Age and Roman Britain. Having chaired the Council of Europe's Cultural Heritage Expert Committee for five years, he also has an interest in the processes of converting damaged landscapes, such as Holocaust camps, into memorial sites.

Professor David Crouch is Professor of Cultural Geography at the University of Derby. David's work, research and writing concerns cultural geography and the complexities of encounters with life, self and the world, especially through processes of encounters with space, and draws upon multi-disciplinary theory on body-practice and performance. This work is exemplified in 'everyday' and 'exotic' performance in complexities through flows of everyday life, mobilities and dwelling, for example in leisure and tourism performance; artwork and art practice; visual culture and distinctive ideological positions. Writing includes these areas and their engagement with such as lay knowledges, landscape, nature, popular culture and new landscape economies. He recently completed 'Flirting with Space: Thinking Landscape Relationally', *Cultural Geographies*, Vol. 17, No. 1, 5–18, and is currently writing *Flirting with Space: Journeys and Creativity*, Ashgate 2010.

Dr Jerome de Groot teaches at the University of Manchester. He is the writer of *Royalist Identities* (Palgrave Macmillan, 2004) and numerous articles on subjects ranging from early modern Royalism to contemporary popular history. His latest book, *Consuming History*, was published by Routledge in 2008.

Dr Roy Jones is a Professor of Geography, Dean of the Centre for Research and Graduate Studies in the Faculty of Humanities and Co-Director of the Sustainable Tourism Centre at Curtin University of Technology in Perth, Western Australia, where he has worked since 1970. Prior to this he obtained degrees in Geography from Sheffield, Newcastle upon Tyne and Manchester Universities. He is an historical geographer with research interests in sustainability and heritage issues. He has been Human Geography Editor of *Geographical Research: Journal of the Institute of Australian Geographers* since 2000.

Dr Tom Mordue is Assistant Dean (Research) at Teesside Business School and has research interests in tourism management and governance, tourism production and consumption, rural and urban policy, spatial management and regulation, and public space. He has published in leading journals including *Environment and Planning*, *Tourist Studies*, *Annals of Tourism Research* and *Leisure Studies*.

Professor Nigel Morgan is Professor of Tourism Studies at UWIC's Welsh Centre for Tourism Research. Nigel has a background in research, policy development and marketing acquired at national agency and local government levels and whilst much of his research focuses on issues of identity and citizenship in tourism, he also has an international reputation in destination marketing. He is currently co-editing *Tourism and Inequality* with Stroma Cole to be published by CABI and the third edition of *Destination Branding* with Annette Pritchard and Roger Pride (Elsevier).

Dr Yaniv Poria is a Senior Lecturer in the Department of Hotel and Tourism Management at Ben-Gurion University of the Negev, Israel. He holds a PhD in Heritage Tourism Management from Surrey University. His main research interest is the management of heritage in tourism.

Professor Annette Pritchard is Professor of Critical Tourism Studies and Director of the Welsh Centre for Tourism Research. She teaches and writes on a range of themes to develop critical interpretations of the interplay between tourism and social structures, experiences and identities. She is particularly interested in social inclusion and gender inequality and leads a number of international research collaborations and supervises several doctoral studies in these areas. She has published ten books and her work has been translated into Chinese, Turkish, Spanish, Italian, Korean and Russian. Her book, *Tourism and Gender: Embodiment, Sensuality and Experience*, was published by CABI in August 2007 and she is currently working on the third edition of *Destination Branding* (Elsevier) and *Tourism, Embodiment and Identity* (Channel View).

Dr Tony Schirato is a Reader in Media Studies at Victoria University, Wellington in New Zealand. He has (co)authored books on academic communication, communication and cultural literacy, Foucault, Bourdieu, globalization, Asian cultural politics, the cultural field of sport, and visual culture, including *Understanding the Visual* (Sage Publications, 2004) with Jen Webb. His most recent book (with Anita Brady) *Understanding Judith Butler* was published by Sage in 2009.

Dr Martin Selby is a Senior Lecturer in Tourism Management at Liverpool John Moores University. Martin has presented conference papers at international conferences, and published journal articles and book chapters in the field of urban tourism, heritage, place image, and the experience of tourism. He has delivered keynote speeches and invited talks for various organizations, including one entitled

'Curly Perms or Culture' at the Institute of Travel and Tourism's Aspire Conference. Martin recently published a sole-authored book entitled *Understanding Urban Tourism: Image, Culture, and Experience* (Tauris, 2003).

Dr Tom Selwyn is Professorial Research Associate of the Department of Anthropology and Sociology at the School of Oriental and African Studies (SOAS, University of London), where he teaches the 'Anthropology of Tourism' at postgraduate and undergraduate levels. His field research in India, Palestine/Israel and the Mediterranean has included caste and ritual, nationalism, landscape, tourism and hospitality, imagery, political symbolism and post-conflict development. He is widely published in these areas. Edited books/special journal issues include *The Tourist Image* (John Wiley and Sons, 1996, with Jeremy Boissevain), *Contesting the Foreshore* (MARE, 2004, with Rachel Radmilli), *Turning Back to the Mediterranean* (Mediterranean Voices, 2005), and *Thinking through Tourism* (Berg, forthcoming, with Julie Scott). Since 1995, he has directed international projects in pilgrimage, tourism and post-conflict development for the European Commission in the Mediterranean and Balkan regions. He is Honorary Librarian and Council member of the Royal Anthropological Institute (RAI) and was awarded the RAI's Lucy Mair Medal in 2009.

Richard Voase holds a lecturing position at the University of Lincoln. His field of interest is consumer culture, specifically the nature of the consumer experience. He speaks regularly at the conferences of the European Sociological Association's Research Network for the Sociology of the Arts, and is the editor of *Tourism in Western Europe* (CABI, 2002) a collection of contributions from nine European countries. His most recent journal article, 'Individualism and the New Tourism: A Perspective on Emulation, Personal Control and Choice' appeared in the *International Journal of Consumer Studies* in 2007.

Dr Emma Waterton holds an RCUK Academic Fellowship in Heritage and Public History at the Research Institute for the Humanities, Keele University. Her research interests centre on examining the discursive constructions of 'heritage' embedded in public policy, with emphasis on how dominant conceptualizations may be drawn upon and utilised to privilege the cultural and social experiences of particular social groups, while actively marginalising others. Further interests include considering community involvement in the management of heritage, exploring the divisions implied between tangible and intangible heritage, and understanding the role played by visual media. Recent publications include the co-authored volume (with Laurajane Smith) *Heritage, Communities and Archaeology* (Duckworth, 2009).

Dr Steve Watson was an Assistant Director of Leisure and Tourism in Local Government before taking up his present post as a Principal Lecturer at York St John University, where he teaches on a variety of tourism and heritage modules. His research interests are in the areas of cultural and heritage tourism and the social, cultural and representational processes by which places are transformed into tourist destinations. He is also concerned with the relationships between heritage and host communities and the nature of the interface between professional practice and community involvement in the formulation and construction of heritage. He has conducted research and published articles in these areas and has recently contributed a chapter on urban heritage tourism to *Reflections on Europe in Transition* (Peter Lang, New York, 2007).

Professor Jen Webb is Professor of Creative Practice and Associate Dean for Research in the Faculty of Arts and Design, University of Canberra, where she also teaches Creative Writing and Cultural Theory. Her academic interests range from neo-Marxist theorising of social practice to aesthetic forms and content, with snippets of semiotics, narrative theory, communication theory and social research along the way. With Tony Schirato she is the author of several books introducing readers to significant cultural theorists or concepts; she is also the author of the short story collection *Ways of Getting By* (Ginninderra Press, 2006) and the poetry collection *Proverbs from Sierra Leone* (Five Islands Press, 2004). Her most recent publication is *Understanding Representation* (Sage Publications, 2009).

Dr Ross Wilson has studied the subject of the archaeology on the Western Front since 2003. His research interests focus on the way in which archaeology might shape and inform the memory of the battlefields. Issues such as the use of images, material culture, embodiment and narrative construction are key features of this work. He is currently working as a post-doctoral researcher on the *1807 Commemorated* project at the University of York.

Acknowledgements

We would like to thank all our contributors for their time and hard work over the period this book has been in preparation and for their patience in responding to our requests for additional information. We are keenly aware that our demands have at times been testing and we are grateful to each of them for providing the myriad detail that goes into making a collection that attempts to reconcile a diversity of viewpoints with a need for some thematic unity and a coherent structure.

The discussions that led to this book took place not only with its active contributors, but also with friends and colleagues over dinner tables, in airport lounges, on trains, on mountain tracks in Rhodes and at various conferences and colloquia that we have attended. Naming names is invidious in omitting all those others who have helped us but it is with intense gratitude that we mention these few. Laurajane Smith has been a constant inspiration, guide and a great friend as we have wended our way through this particular voyage of discovery. Others who have been generous in reading and commenting on our own material include Margaret Atherden and Michael Hopkinson of the 'Culture and Environment in the Eastern Mediterranean' project at York St John University, Gary Campbell, Keith Emerick, Kalliopi Fouseki (for helping with matters Greek) and Aleks McClain (for doing amazing and mysterious things with a computer to make some of the images usable). Thanks also to Val Rose, our Commissioning Editor at Ashgate, for her advice and support.

The editors and publisher gratefully acknowledge the permission granted to reproduce the copyright material in this book. Every effort has been made to trace copyright holders and to obtain their permission for the use of copyright material. The publisher would be grateful if notified of any corrections that should be incorporated into future reprints or editions of this book.

As for errors and omissions, we claim them entirely for ourselves.

List of Abbreviations

AHC	Australian Heritage Commission
AHD	Authorized Heritage Discourse
BT	British Telecom
CBA	Council for British Archaeology
CDA	Critical Discourse Analysis
DCMS	Department of Culture, Media and Sport
DIY	Do-it-Yourself
EH	English Heritage
IDF	Ideological Discursive Formation
MECA	Museum of Ephemeral Cultural Artefacts
MVP	Museum Video Podcast Player
NOMIS	Official Labour Market Statistics
PC	Personal Computer
RSS	Really Simple Syndication
UCLA	University of California, Los Angeles
UNESCO	United Nations Educational, Scientific and Cultural Organization
VR	Virtual Reality
WTO	World Tourism Organization
YAT	York Archaeological Trust

Chapter 1
Introduction: A Visual Heritage

Steve Watson and Emma Waterton

Our turn to representation in this volume is not an isolated event; we are by no means the first to tackle the subject and it would be disingenuous to suggest otherwise. Stuart Hall's edited volume *Representation: Cultural Representations and Signifying Practices*, first published in 1997, was, for example, one of our starting points for this project and remains at its core. We therefore position this volume within the context of a long-established history in sociology that takes representation as a key moment of meaning-making (Chaplin 1994; du Gay et al. 1997; Hall 1997; Evans and Hall 1999; Emmison and Smith 2000; Morra and Smith 2006). This history has wended its way through cultural studies, tourism studies, cultural geography, art history, communication studies, archaeology and anthropology amongst others (Daniels and Cosgrove 1988; Foster 1988; Urry 1990; Bryson et al. 1994; Rojek 1993, 1997; Sturken and Cartwright 2000; Mitchell 2002; Crouch and Lübbren 2003; Schirato and Webb 2004; Smiles and Moser 2005; Rose 2007), and has begun to permeate the research of scholars working in the field of heritage studies, resulting in a smattering of publications that play with ideas of visuality, imagery, visual culture and what can broadly be termed as representational practice. Yet, while representation has long been accepted as a core component falling within the remit of heritage studies, not least because of the development of museology and interpretation studies (Merriman 1991; Hooper-Greenhill 1995, 2000) and critical studies of heritage itself (Walsh 1992; Brett 1996), few attempts to make it the dedicated focus of scholarship have been made. Our purpose with this volume is to grant it that right of entry and raise questions about its relative significance, modalities and various ways of applying and understanding it.

The title of this book is deceptively – and deliberately – simple: *Culture, Heritage and Representation.* In it, we explore some of the ways in which the past has been constituted in the present, with visual culture highlighted as a key medium for communicating and understanding it. Each contribution grapples with this intersection in its own way: some are theoretical and speculative, others comparative and suggestive, and some still are rooted in fine-grained analyses of distinct case-studies. This combination allows the volume to speak to a broad readership, addressing the kinds of issues we believe a range of scholars will be interested in – performativity, hyper-reality, virtuality, non-representational theory and popular memory – whilst also detailing the intricacies of case studies ranging from Greece, the United States, the United Kingdom, Israel and the Western

Front. Throughout all of these contributions there are facets and insights that demonstrate the visual as an *active agent* either in representations or in subjective and inter-subjective responses. Despite their differences in approach, content and background, we think we have something that collectively builds important connections and has the potential to make serious contributions to wider debates regarding representation and visuality. Ultimately, however, we leave it to the reader to negotiate such meanings and feed these contributions into the purposeful advances already unfolding in the wider study of representation.

Heritage and the Visual

The processes that constitute meaning, that frame, reveal and construct the past that we see around us, are essentially visual. Our connections with the past are largely tangible, or have a materiality upon which they depend that makes them objects of heritage, and it is visual culture that lends these objects the means of representation and achievement of meaning. The artistic tradition of the flamenco, for example, could be seen as intangible but its essential visuality is based on a vivid materiality: the body and its movement, the costume and the setting. More typically, heritage has been concerned with the object-artefact. This tangibility – the traditional centrality of material culture, the exhibit, the building, the place, the landscape – has created the sight-nexus of heritage, setting artificial constraints on the ways that heritage can be and *is* perceived and studied. Aesthetes and experts, connoisseurs and curators have thus made heritage their own resort, and their associated skills in interpretation, presentation and representation have defined a dominant discourse that is both powerful and resilient. There are two problems with this discourse, however. The first is its obsession with material culture. This sees the reification of the social relations that create, sustain and reproduce heritage objects as autonomous things that tell their own story about the past, which is expressed, limited and satisfied by their very materiality. This is noted, for example, by Dubbini in his examination of the visuality of landscape, where he proposes:

> [A] history of 'modes of seeing' rather than a history of images, a history based on perceptive phenomena, scientific acquisitions, mental equipment, the epistemological framework, and the personal and collective visions that influence representations (2002, 8–9).

The second problem is that the reification of heritage has encouraged scholars to be equally focussed on materiality and its associated representation practices. To be sure, this is usually in the service of deconstruction and critical analysis, but it has arguably led to a focus on representations and representational practices rather than the dynamics that might be investigated in the spaces between representation and response.

In addition to identifying problems germane to the broader analysis of representations, this volume also challenges some of the specific, underlying assumptions and precepts upon which understandings of *heritage* are built. Using emerging theories and ideas about the visual and the process of subjective engagement, chapters within the volume address a notable shortcoming in the existing literature in addressing heritage not as a set of objects, but as a process involving the construction of meaning (after Smith 2006). The result is a lively challenge to the discursive reproduction of accepted ways of thinking about heritage, particularly those that sustain differential access to social power and thus the means to define and encounter 'heritage'. Subsequently, 'meaning' for many heritage scholars is no longer seen as constructed primarily from the perspective of dominant groups, although this debate, we believe, has a long way to go. The relationship between the construction of meaning and the visual culture upon which it seems to largely depend is thus the likely key figure to emerge in this book, which attempts to chart a new course using the visual as a framework for analysis. Hopefully, this collection will add to the valuable, *critical* work already developing in the field of heritage studies.

We are aware, however, that in charting this course we are entering, sometimes tentatively, areas that perhaps at first do not appear linked and emerge out of varied disciplines. Thankfully, others have opened the way before us. David Crouch and Nina Lübbren (2003, 1), for example, addressed the heterogeneity of such enquiry, carefully leaving questions open for further discussion in their collection *Visual Culture and Tourism*. Laurajane Smith, in her work *Uses of Heritage* (2006), effectively relocates heritage away from its crude delineations of object-orientation, inherent value and reification, replacing this with the idea of heritage as an essentially cultural process. As well, Jen Webb revisits the idea of representation as culturally mediated in her recent publication *Understanding Representation* (2009), where she interrogates representations that appear natural, revealing them to be products of discourse. There have been others, a number of which have been at pains to point out the significance of the visual, not least because of its primacy in tourism (MacCannell 1999; Urry 1990; Lash and Urry 1994; Morgan and Pritchard 1998; Crang 1999; Selby 2004). Others, still, have been exercised by its centrality to the consumption of heritage sites and objects through interpretation, display and their distinctly social and constructivist frameworks (Palmer 1999; Hooper-Greenhill 2000; Kirshenblatt-Gimblett 1998). David Brett (1996) has traced the visuality of heritage over the period of centuries, arguing that it has become a form of popular history. This in turn is inscribed with a 'visual ideology', a term coined by Hadjinicolaou (1978, cited Brett 1996, 7) to describe the ways in which art works contain the major thematic elements of a dominant ideology. For Brett, heritage itself is inscribed in this way:

> Buildings, parks, exhibitions and displays are created by organisations that have their particular values and assumptions inscribed in their products. These products are prescriptive and normative because they have been given concrete

> form; their guiding intentions (conscious or otherwise) can often be quite precisely assessed. A direct study of the physical manifestations of heritage – quite literally, its construction – reveals something of the values and ideological functions of the concept (1996, 12).

We are also well aware that many readers will have encountered recent debates surrounding issues of visuality, visual design, representation, the touristic image and so forth for some time now, with notable contributions from Rorty (1980), Bryson et al. (1994), Jenks (1995) and Selwyn (1996). Those familiar with social semiotics will have consulted the work of Gunther Kress and Theo van Leeuwen (2006) *Reading Images: The Grammar of Visual Design*, for example, or Gillian Rose (2007) *Visual Methodologies: An Introduction to the Interpretation of Visual Material*, for explorations of visuality and anthropology. A certain distance has already been travelled for us, with considerable material dealing with the power of images and their ability to construct, produce, communicate and constitute the world in meaning in a full range of social settings. It is this understanding of representations that we adopt in this volume. Representations, from this perspective, thus become far more than pictures alone, allowing for the analysis of ways of acting, identifying and being, as well. Allowing 'representations' to stretch beyond discrete semiotic features in this way enables analysts to make links with the wider social contexts and practices from which they draw (and give) meaning. Given the breadth of analysis that has emerged over the last 20 years, it should come as no surprise, then, that the contributors to this volume each provide their own views of the relevant literature and we have no intention of making arbitrary decisions that will pull those complexities into a simple unified approach. Rather, our aim is to showcase the richness of current thinking on the subject of representation and the visual.

To find new ground in an existing debate is always a challenge. In this volume, we are using the visual to explore both representation and response, keenly aware that these are uneven relationships and unequal engagements. The power of representation lies with few, yet the subjective response is owned by many. But what happens in between, and how can an understanding of visual culture help us to understand these engagements? What does the act of representation draw upon, and why? What is represented, and how? Is the process of encoding contingent on inter-subjective understandings and meaning? Is the object ever autonomous or always embedded in one understanding or another? And so we engage politics, dissonance and the essence of culture as a process of knowledge production that is at once concrete and prefigured yet also challenged, negotiated, provisional and subjective. Applied to heritage, such analysis yields much to illuminate Smith's (2006, 3; Smith and Waterton 2009) construction of heritage as a process, linked as it is to memory, identity, politics, place, dissonance and performance. That said, there is nothing inherently new about examining the social world of the subjective engagement outside of heritage studies. Judith Adler (1989), for example, in focussing attention on the traveller and the tourist, was an early exponent of the search for the subjective voice that animates the view from the subject and the process of 'self-fashioning' (1989, 1368).

Just as these ideas are as yet only loosely framed and not yet fully articulated, so the methods of analysis are in their infancy, at least in this field. Decentring the object of heritage meant focussing attention on the processes of objectification and the disciplinary and institutional power relations that held them in place. Simultaneously, the idea of 'subjectivity' has required deeper understandings than could be provided by the ubiquitous visitor survey. Tourism of necessity generates text; more so now with the internet and its use both as a medium for promoting destinations and as a means of interactive engagement and commentary. The speed and immediacy of that textual creation challenges the existence of prefigured narratives or at least their mode of construction. Similarly, the moment of engagement can be seen as an act of subjective performance in which meaning is constituted, or re-constituted, in a nexus of the subject's own making. The application of various forms of textual analysis within the intertextuality of tourism and heritage, and the value of specific methodologies in discourse analysis, are helping to reveal and elucidate these newly centred processes.

The Themes of This Book

Perhaps the clearest objective to emerge from our Introduction so far is the impetus for new debate, the beginnings of which are presented here. It is probably apparent, however, that there is considerable variety in the way that this debate may unfold, and for this reason we have divided the subsequent chapters into those dealing with Relocating the Visual (Part I), Representation and Substitution (Part II), Visual Culture and Heritage Tourism (Part III) and Constructing Place (Part IV). These four parts are structured in a way we hope ignites discussion. Such divisions, however, are immediately invidious, inviting criticism, alternative categories and classifications. This we accept. The sections are there to provide a semblance of structure and thematic development in the full acknowledgement that it may have been done just as well another way.

Part I: Relocating the Visual

The chapters contained in Part I provide what we consider to be an essential and accessible theoretical orientation: understanding the visual in relation to the social and (re)locating it as a social and cultural process. The implications of this are both compelling and challenging. If the visual is a social process, and thus more than just a way of seeing, how might it be understood socially? How does it relate to cultural narratives? How does it differ from linguistic or written encodings of meaning, if at all? What does it reveal, and what does it obscure? The purpose of this first section of the volume is thus to gently introduce the more unfamiliar readers to a range of key theoretical debates and concepts for analysing the visual without compromising their complexity, and thus utility, for the more seasoned scholar. In Chapter 2, Tony Schirato and Jen Webb capture the essence of historical debates surrounding the visual in an account of the conceptual trajectory of the image from a knowledge-bearing signifier to a story, or part of story. This they explore in terms of urban space, with its social and economic constraints, and the visual juxtaposition that creates narrative beyond the essentially linguistic structure of semiotics. The resulting 'narrativising' of visual material can then be read and understood as a constitutive regime that stands in relation to the society in which it exists.

In Chapter 3, Martin Selby offers a reflexive account of the active nature of cultural heritage consumption through the lenses provided by phenomenology. His discussion draws upon themes of visualizing, representing, performing, perceiving, knowing and acting in a challenging attempt to construct and contest many of the concepts that endure within the wider sociological literature. His aim is to break down dichotomies between representational and non-representational theories and explore how heritage is experienced in as full a sense as possible. Where, for example, does representational practice end and a subjective response begin in terms of the construction of meaning? What part does the visual play in this debate? The approach Selby offers is cautious of bending heritage studies into an either/or debate, focussing instead upon the nuances of everyday experience that meld the visual, discursive and performative. We shall see throughout this book that Selby is not alone in his concerns; this is a conceptual problem that exercises much of the recent literature, led particularly by work emerging from tourism studies.

In the third of our introductory chapters, Chapter 4, David Crouch's theoretical exploration of the visual attempts to dislocate it from the systematized and privileged position it has occupied in cultural theory in general and heritage studies in particular. Whilst acknowledging the visual as a key component in the construction of heritage meanings, he discusses it within a context of 'comingling energies' that constitute it in moments of subjective engagement. Such moments are, in turn, characterised and animated by the performativity of both heritage production and the engaged subject. This nexus of constitutive engagement renders heritage as *emergent* as well as represented and the implications of this for the visual are that non-representational narratives can assume significance over an above what is merely represented. This broader perspective allows for a fluid and dynamic relationship between subject and heritage object that projects heritage beyond the familiar debates of the last 20 years into more 'nuanced personally varied issues' and a 'gentle politics that emerges from the quieter affects of people coming to their own heritage' (Crouch, this volume).

Part II: Representation and Substitution

Part II brings together what on first appearance may seem an eclectic collection of chapters. They are united, however, in their critical unpacking of many of the issues that orbit representation, such as identity, memory and reality, honing in particularly – for some explicitly and others implicitly – on the idea of substitution. For Ross Wilson (Chapter 5), the image can be an active agent in constituting popular memory, through which a particular understanding of the Western Front has come to stand in for the many conflicts of the First World War. In his study of visuality, Wilson argues that collective memory is dominated by the duality of bleak images of trench warfare and the contrasting tranquillity of the war cemeteries of Northern France and Belgium. In visually constituting a popular memory of the Western Front, the stoicism of the common soldier in the face of unspeakable suffering in a pointless war is tightly crafted into contemporary understandings of that past. The link between the image and remembering is central here, not just in shaping and informing concepts of British nationalism and affirming a particular memory of the part that emerges from this, but also in alternative memories wherein the role of the archaeological image can play a vital part. This is particularly important where revisionist historians have sought to disrupt abiding popular memories and substitute a narrative that centralizes military history and emphasizes the eventual allied victory. Archaeological imagery, however, can not only revivify popular memory in the face of revisionist history, it can offer a wider basis beyond nationalistic concerns to form the basis for a shared European memory of these events.

Equally implicated in the role of substitution are the new digital technologies available to museums. Jerome de Groot's (Chapter 6) survey of the new virtuality of museum displays, and the context this provides for new performances of museum-user engagement, is illustrated with reference to the *Tate Online*. For de Groot, virtuality 'provokes a fundamental shift' in the concept of performativity and the ways that heritage itself might be performed and interpreted. The virtual museum not only provides content, therefore, but new forms of engagement in which a new visuality is implicated in processes of searching, selection and interactivity. It is not only the virtuality that is new, but also, inevitably, the modalities of display, which online engenders a purer encounter with the object without distraction or interruption. The main implication of this is that the museum enters a broader community of representation, a 'folksonomy', in which the authority and legitimacy of the museum itself is challenged, its visual tropes fractured and destroyed by an ever more active and interactive community.

The theme of substitution is continued by Richard Voase (Chapter 7) in his survey of the emergence of linkages and interplays between the hyper-real, as conceived by Baudrillard in his influential *Simulcra and Simulation* (1981), and visualizations of the past; or, to put it more simply, between representation and reality. Although not solely a product of recent technological advances, the format Voase explores, cinema, remains relevant in its ability to amplify and intensify different versions of reality in historical settings. Hyper-reality in these contexts becomes a 'beyond real' representational and image system, and when it is applied to heritage it adds a further dimension to the already uneasy relationship between heritage and conventional historicity. History in this formal sense is thus compromised, with hyper-reality intimating pasts that there bear little historical scrutiny. Using Baudrillard's *four phases of the image*, by which the hyper-real can be evaluated in specific instances, and recalling Crouch's concept of heritage as dynamically emergent, Voase explores the extent to which individual meanings are continually authored in a way that replaces both history and its imitators in the visual cultures of hyper-reality.

Part III: Visual Culture and Heritage Tourism

It is, of course, the spectacle-creativity of tourism that provides for much of the presentation and consumption of heritage. As Allcock has expressed it:

> '[H]eritage' is not just that which has come down to us from the past: it is one version of that past, which potentially competes with other possible versions, but which has come to be sponsored as appropriate and acceptable. Typically the ruling considerations which shape the construction of the past as 'heritage' in this way have to do with the perceived demands of the tourism industry (1995, 100–1).

It seems to us entirely appropriate, therefore, that Part III should include chapters that deal primarily with heritage tourism. Crouch and Lübbren (2003, 12; see also Crouch 1999) make the point that tourism is also implicated in the production of visual culture as well as in its consumption, with the tourist thus involved in an *encounter* not just an act of consumption. This encounter generates its own visual culture, from taking photographs and sending postcards, to collecting and assimilating texts of various kinds that add meaning to the encounter. The intersection of visual culture with tourism also draws in processes of commoditization and exchange, on a global scale, both inalienable dimensions of the contemporary world (Schirato and Webb 2004). For Meethan, the analysis can be applied to the issue of touristic significance as follows:

> Within the restrictions of the global economy, policies and marketing strategies assign symbolic and aesthetic value to the material attributes of space. In turn these representations or narratives of people and place assume an exchange value as the objects of consumption becoming commodities to be traded and consumed the same way as the material goods and services which are associated with them (Meethan 2001, 37).

Meethan's analysis places representational practices at the core of the process by which space becomes touristic, particularly through 'civic' marketing strategies, the significance of which is well established (see Light and Prentice 1994). Graham et al. (2000, 163–7) provide a closer account of this mechanism in discussing the significance of 'civic consciousness' as a locus for collecting and representing images of place for both tourists and citizens, and the importance of heritage as a component in this process. National organizations such as English Heritage and local and regional organizations including local authorities and regional tourism bodies for whom tourism is a key driver of economic growth are central to this process, generating text and images from embedded interpretation to brochures, leaflets and websites, all in the service of attracting more visitors and representing a particular discourse.

Collectively, the chapters in this Part offer a balance of theoretical issues and empirical work, illustrated through detailed case studies such as the beach holiday, 'celebrity' sites associated with Robin Hood and Brother Cadfael, the imagery of 'Englishness', the historic city of York and promotional literature for a range of organizations associated with tourism. It begins with Annette Pritchard and Nigel Morgan's (Chapter 8) study of the beach holiday, a form of tourism with its own particular heritage and visuality. In one sense, it is 'natural heritage' liminally placed between land and sea where quotidian norms are replaced by display, performance and transgression (see Shields 1991, 83–101 for an account of the theoretical antecedents). This, in turn, as Pritchard and Morgan demonstrate, has initiated and sustained a 'gendered visual rhetoric' that in its overwhelmingly sexual and implicitly sexist practices continues to frame a tourist experience that actively facilitates the objectification of women. This *carnivalesque*, originally conceived by Bahtkin, recalls the colour and animation of medieval holidays where performance and display constitute and reconstitute meaning and personal identity in ways that are broadly reflective of the fluidities and constitutive dynamism of heritage engagements. Pritchard and Morgan impart a clear understanding of how 'the beach' is created and remade into a tourist destination for the white, male, Western tourist, and provide important discussions of how this visuality implicates issues of power, politics and gender.

Where Pritchard and Morgan have identified a clear continuity with the essentially visual and performative cultures of the past, Roy Jones (Chapter 9), by contrast, identifies and locates a series of literary vignettes abstracted, preserved and represented through contemporary tourist representations. Here, tourist destinations are visualized through the contemporary media of film, television and the internet, with a visuality based essentially on the formulae of television drama. This 'hot authenticity' of touristic make-believe, as Selwyn (1996) termed it, reframes places as reference points in television drama and, using the Brother Cadfael stories and Robin Hood and their associations with Shropshire and Nottingham in the UK, he identifies a particular engagement of contemporary heritage tourism with the visual and dramatic qualities of the medieval as previously described by Umberto Eco (1986). This symbiotic relationship between tourism and the media industries has resulted in the appropriation of many meanings by destinations so as to create a focus for particular types of heritage tourism closely linked with economic regeneration.

For Emma Waterton (Chapter 10), such touristic moments lend visibility as much to the present as the past, invented or otherwise. The visual in heritage tourism is thus a means 'to address real issues in the political, social and cultural arenas', recalling Allcock's persuasive account of heritage as a political concept (1995, 100–1). The images that are used to represent, promote and ultimately construct heritage experiences are fundamentally the cultural symbols of an elite social group. They are presented, however, as a 'consensual' past, a past without conflict and a past that is more to do with leisure, entertainment and family fun than with historical verities and continuities. The ideological implications of this provide for a fruitful analysis of the role of the visual in perpetuating existing power-relations, presented through the lenses of discursive and rhetorical approaches to understanding narratives in the imagery of official guides and brochures. On this basis, Waterton identifies a specific ideological discursive formation (IDF) that has come to dominate the framing and presentation of heritage and heritage tourism, in England at least. This is an example in practice of the *authorised heritage discourse* identified by Smith (2006), through which the objects of heritage, specifically the symbols of an elite discursive dominance largely focused on national identity, come to be expressed through aesthetic judgements and animated by sites of remembrance, commemoration and monumentality. Heritage is thus a process of selection, representation and closure, by which the 'communal national heritage is effectively "completed" … at the expense of alternative understandings of heritage'. Indeed, for Waterton, there is no evidence in official representations of the need to deconstruct the conventional visual imagery used to depict heritage, and despite stated objectives about inclusivity, a heritage is constructed that is laden with its own singularity of meaning and the performative power to sustain and reproduce itself.

In the context of what might be described as the new urban tourism, culture and heritage have figured largely in attempts to commodify space and signify post-industrial regeneration. Tom Mordue (Chapter 11) examines these processes in the context of the interface between global and local economic change. Here, local governance acts as the mediating institution whose decisions will most clearly translate that engagement onto the re-configuration of space in a way that is effectively a consumerist/promotional model of urban renaissance. Mordue directs his attention to the city of York in northern England, where a ready supply of ancient buildings, a rich and popular genealogy and a historic urban morphology have endowed it with a long-standing status as a destination for both domestic and international tourists. For Mordue, however, there are implications for the city as a performative arena where touristic and local meanings converge and conflict depending on the social differentiations that are apparent among and between visitors and residents. Such schismatic lines have been obscured, however, by emergent urban governance that privileges, through its scope and private sector methodologies, a model of both residents and visitors as consumers rather than citizens. The construction of public-private sector partnerships to oversee tourism has exacerbated this process of commodification as cities seek to compete in the global tourism market. The distinct visuality that cities such as York then assume is thus a spectacularized attraction rather than social entity conditioned by democratic decision-making and an inclusive polity. The commercial and promotional representation of the city displaces and eventually replaces existing local meanings and this spills out into the ways in which space is differentiated, visualized and produced.

In the final chapter for this part, Tom Selwyn introduces us to the dialogues between image-makers and narratives of the self exercised through the medium of 18 promotional images drawn from a range of geographical contexts. In setting the parameters of his analysis, Selwyn introduces us to the concepts of the body, collectivity and exchange, all three of which provide a means for organizing, negotiating and experimenting with the 'self'. At the same time, Selwyn argues, these myriad possibilities converge on a single storyline, as when taken together, the 18 images under analysis provide also a '... focussed summary of the modern world system of liberal, democratic, states underpinning a global regime of free markets in which we, as tourists, may roam as free consumers'. In doing so like 'jugglers in a hall of mirrors' they construct a self in relation to the image laden world around them, a world that contemporary tourism richly facilitates. Brochures thus become the philosophical treatises offering complex arrays of identity, identification and position, both spatially and temporally.

Part IV: Constructing Place

Visuality is created, encountered and regulated in 'place'; literally and figuratively. As Part III identifies, tourism is implicated in the construction of place *literally* through its physical developments and infrastructures, and through the representative and performative accretions of place imagery which then differentiate destinations according to their market position and their 'target audiences'. Urry, for example has demonstrated that it was visitors who produced the Lake District (United Kingdom) as it is conventionally known and understood, so that it was transformed from the barren landscape of the seventeenth century to the artistic scenery of later centuries (1997, 198). Such tourist places are also, however, replete with meaning: for those who live there, those who visit and for those who would represent and prepare such places as attractions. The visual thus becomes active in context, in a place where meanings emerge in moments of representation, performance engagement and interaction, as described by Mordue in the city of York. Representations and visual imagery are thus intimately embedded in the mechanisms that construct and *make real* a range of imagined, historic and/or mythical places. Shields (1991) has explored these processes at their most intense in *Places on the Margin*: 'Sites are never simply locations. Rather they are sites for someone and of something' (1991, 6). Lefebvre (1994) asserted that places (or spaces) were not mere crucibles or containers of activity but were in some way expressions of that action and social momentum, and of the social processes that filled, animated and ultimately defined them. The users of spaces are no small part of this process. Indeed, Lefebvre made their actions and understandings one of his triad meanings that generate the production of space. This idea of place draws together some of the significant strands in this book and forms the basis for our third and final section.

Yaniv Poria (Chapter 13) opens this section on the construction of place with an examination of the user's perspective on the provision of display at heritage sites. This shift in perspective necessarily involves a movement of emphasis away from what is displayed to what is expected, sought and ultimately experienced by the visitor and how this experience comes to be configured in a particular way. In short, representations become less significant than the ordered meanings of expectation and belief. As a result, heritage becomes an abstraction based on perceptions rather than inherent value. For Poria, this sense of heritage may come to strengthen the differences and potential conflicts between people and cultures rather than resolve them. Much, however, depends on the nature of the tourist and their differential expectations and motivations, and the various ways in which they respond to heritage displays. For example, for the group termed by Poria to be 'identity builders', the visual supports collective memory and validates belief in myth and origin rather than serving any conventionally objective purpose in interpretation. These motivations differ in subtle and complex ways from those attributed to Poria's other categories of the 'multicultural minded audience' and the 'ticking and guilt-reducers'. Either way, heritage becomes separated from its objects by the motivations of those for whom it has meaning, generating the possibility that what might be viewed are historical objects without heritage significance and heritage productions without objects

Tim Copeland's (Chapter 14) analysis investigates the visual in the construction of heritage places where visible, material remains are, paradoxically, often lacking or at best characterized by 'low visibility'. Roman remains in the United Kingdom are often no more than traces in the landscape, revealed by aerial photography. Visible ruins are rare, and when sites are excavated often all that remains are foundations and footings, parts of the building that the Romans themselves would not have seen. The result is often a strange mix of image, myth, text and material culture in a landscape context bound together by that great landscape marker – the roman road. The complex visuality of Roman heritage encourages visitors to adopt a variety of 'visual roles' that enable the remains and traces to be 'viewed'. Often, this results in what MacCannell (1999) referred to as marker involvement, where in the absence of an object, the interpretation media provide a substitute to enable the visitor and sightseer to be there in such a role. This supports Poria's notion of heritage occurring whether or not there is an object available to view; it depends instead on whether or not there is a motivation to seek it and give it meaning. In Copeland's study, the needs of different audiences are also apparent in what, and how, the traces are presented, especially where scholarship and popular understandings converge. The interconnectedness and fluidity of effective visually-based interpretation in the landscape is the real opportunity here, thawing the frozen monument from its sterile vestigial state by using visual means to add value and meaning.

A number of authors in this book, including Poria and Copeland, discuss the ways in which visuality drifts away from the objects of heritage, becoming part of the process by which heritage is constructed and understood. For Steve Watson (Chapter 15), this process is complete when visuality assumes the status of metaphor describing the social and cultural significance of certain constructed pasts. This social visuality provides a method for differentiating the value placed on a variety of heritage objects that are then ordered according to their significance within a particular cultural domain such as heritage tourism in a specific destination. This ordering, the consequence of a distinctive *scopic regime*, may in turn lead to the addition of further visual value in the way that objects, places and monuments are selected, framed, represented and managed for the use of visitors. In his study of visuality in the heritage tourism of Rhodes, Watson identifies a series of distinct affinities in the island's heritage, which are revealed by the visuality of its heritage and which ultimately seek to reveal and emphasize its Greekness in relation to the identity politics of the island's history.

Culture, Heritage and Representation

Common themes are hard to find in such multidisciplinary and interdisciplinary contributions, but that, in a sense, is the purpose of this book. There is, however, a key thematic element that is wrapped around the issue of the visual in culture and heritage representations that perhaps deserves a last mention before we turn things over to the contributors. Visuality is strongly implicated in the processes that contemporary theory associates with heritage. The visual is thus either centrally or partially involved (depending on your standpoint) in the production of heritage, heritage tourism and in the nature of the engagement between the objects of these and the active subject. The construction of meaning in these engagements is thus variously dependent on the power of representation and the involvement of the subject in the constitutive moment. Fluidity and dynamism may characterize such encounters but at some point meaning is captured and expressed in the modalities of visual culture and representational practice. We do not seek to simplify these complexities, but rather to display them.

In an object lesson in visuality and meaning, Velazquez both obscures and represents the King and Queen of Spain in his painting *Las Meninas*, and without revealing *them* directly, reveals their dominance, their centrality and power (Foucault 1970; Hall 1997, 56–61). Velazquez's painting demonstrates one key feature of visuality that is clear throughout this book, which is that it has a dynamic quality in the construction of meaning, and that rather than presenting windows on the past, it can obscure as much as it displays: things that are absent being as important as those that are present, until significations are understood and cultural processes are revealed. When this occurs, the role of the visual in reification and in naturalizing social relations, especially when it is so closely interwoven with the representation and reception of heritage, is made evident, not merely in representational terms or as a signifier, but as an instrument of cultural power, active and operational in its contexts (see Mitchell 2002, 1–2 for a discussion of this in relation to landscape).

In short, we are concerned here with making sense of the visual and the way that the visual makes sense of the world and, more particularly, the past. Where and how is meaning constructed? Where does power lie? How is it represented and what is the nature of the dynamic in relation to the subject? What this means for the significance of visual culture we have left for the reader to decide, once they have considered the variety of contexts and viewpoints presented in this book. It seems clear, however, that whether in the depths of history or the heat of technological change, the visual maintains at least some of its significance in ordering and sustaining our views of the world and our knowledge of its past. Although there is inevitably more work to be done to make fuller sense of the dialogues between culture, heritage and representations, we hope the chapters collected together in this volume stimulate enough interest to continue the debate.

References

Adler, J. (1989), 'Origins of Sightseeing', *Annals of Tourism Research*, 16, 7–29.

Allcock, J.B. (1994), 'International Tourism and the Appropriation of History in the Balkans', in M.F. Lanfant, J.B. Allcock and E.M. Bruner, (eds), *International Tourism: Identity and Change*, 100–12, (London: Sage Publications).

Baudrillard, J. (1981), *Simulacra and Simulation: The Body in Theory*. (Ann Arbor, MI: University of Michigan Press).

Brett, D. (1996), *The Construction of Heritage*. (Cork: Cork University Press).

Bryson, N., M.A. Holly and K.I. Moxey (eds) (1994), *Visual Culture: Images and Interpretations.* (London: Wesleyan University).

Chaplin, E. (1994), *Sociology and Visual Representation*. (London: Routledge).

Crang, M. (1999), 'Knowing, Tourism and Practices of Vision', in D. Crouch (ed.), *Leisure/Tourism Geographies: Practices and Geographical Knowledge*, 238–56, (London: Routledge).

Crouch, D. (1999), 'Introduction: Encounters in Leisure and Tourism', in D. Crouch (ed.), *Leisure/Tourism Geographies: Practices and Geographical Knowledge*, 1–16, (London: Routledge).

Crouch, D. and N. Lübbren (eds) (2003), *Visual Culture and Tourism*. (Oxford: Berg Publications).

Daniels, S. and D. Cosgrove (1988), *The Iconography of Landscape*. (Cambridge: Cambridge University Press).

Dubbini, R. (2002), *Geography of the Gaze: Urban and Rural Vision in Early Modern Europe*, trans. L.G. Cochrane. (Chicago: University of Chicago Press).

du Gay, P., S. Hall, L. Janes, H. Mackay and K. Negus (eds) (1997), *Doing Cultural Studies: The Story of the Sony Walkman.* (Milton Keynes: The Open University Press).

Eco, U. (1986), *Travels in Hyperreality: Essays.* (San Diego: Harcourt Brace Jovanovich).

Emmison, M. and P. Smith (2000), *Researching the Visual.* (London: Sage Publications).

Evans, J. and S. Hall (1999), *Visual Culture: A Reader.* (London: Sage Publications).

Foster, H. (ed.) (1988), *Vision and Visuality.* (New York: The New Press).

Foucault, M. (1970), *The Order of Things.* (London: Tavisock).

Graham, B., G.J. Ashworth and J.E. Tunbridge (2000), *A Geography of Heritage, Power, Culture and Economy.* (London: Arnold).

Hall, S. (ed.) (1997), *Representation: Cultural Representations and Signifying Practices.* (London: Sage Publications).

Hooper-Greenhill, E. (1995), *Museum, Media, Message.* (London: Routledge).

Hooper-Greenhill, E. (2000), *Museums and the Interpretation of Visual Culture.* (London: Routledge).

Jenks, C. (1995), *Visual Culture.* (London: Routledge).

Kirshenblatt-Gimblett, B. (1998), *Destination Culture: Tourism, Museums and Heritage.* (Berkeley, CA: University of California Press).

Kress, G. and T. van Leeuwen (2006), *Reading Images: The Grammar of Visual Design.* (London: Routledge).

Lash, S. and J. Urry (1994), *Economies of Signs and Space.* (London: Sage Publications).

Lefebvre, H. (1991), *The Production of Space.* (Oxford: Blackwell).

Light, D. and R.C. Prentice (1994), 'Who Consumes the Heritage Product? Implications for European Heritage Tourism', in G.J. Ashworth, and P.J. Larkham, (eds), *Building a New Heritage, Tourism, Culture and Identity in the New Europe*, 90–114, (Routledge: London).

MacCannell, D. (1999), *The Tourist: A New Theory of the Leisure Class.* (Berkeley, CA: University of California Press).

Meethan, K. (2001), *Tourism in Global Society, Place, Culture, Consumption.* (Basingstoke: Palgrave).

Merriman, N. (1991), *Beyond the Glass Case: The Past, the Heritage and the Public in Britain.* (Leicester: Leicester University Press).

Mitchell, W.J.T. (2002), *Landscape and Power.* (Chicago: University of Chicago Press).

Morgan, N. and A. Pritchard (1998), *Tourism Promotion and Power: Creating Image, Creating Identities.* (Chichester: Wiley).

Morra, J. and M. Smith (2006), *Visual Culture: Critical Concepts in Media and Cultural Studies.* (Abingdon: Routledge).

Palmer, C. (1999), 'Tourism and Symbols of Identity', *Tourism Management*, 20, 313–21.

Rojek, C. (1993), *Ways of Escape: Modern Transformations in Transport and Travel*. (Basingstoke: Macmillan).
Rojek, C. (1997), 'Indexing, Dragging and the Social Construction of Tourist Sights', in C. Rojek and J. Urry, (eds) (1997), *Touring Cultures, Transformations of Travel and Theory*, 52–74, (London: Sage Publications).
Rorty, R. (1980), *Sensuous Geographies*. (London: Routledge).
Rose, G. (2007), *Visual Methodologies: An Introduction to the Interpretation of Visual Materials*. (London: Sage Publications).
Schirato, T. and J. Webb (2004), *Understanding the Visual*. (London: Sage Publications).
Selby, M. (2004), *Understanding Urban Tourism: Image, Culture and Experience*. (London: I.B. Tauris).
Selwyn, T. (ed.) (1996), *The Tourist Image*. (Chichester: Wiley).
Shields, R. (1991), *Places on the Margin: Alternative Geographies of Modernity*. (London: Routledge).
Smiles, S. and S. Moser (2005), *Envisioning the Past: Archaeology and the Image*. (Malden, MA: Blackwell Press).
Smith, L. (2006), *Uses of Heritage*. (London: Routledge).
Smith, L. and E. Waterton (2009), *Heritage, Communities and Archaeology*. (London: Duckworth).
Sturken, M. and L. Cartwright (2000), *Practices of Looking: An Introduction to Visual Culture*. (Oxford: Oxford University Press).
Urry, J. (1990), *The Tourist Gaze: Leisure and Travel in Contemporary Society*. (London: Sage Publications).
Urry, J. (1997), *Consuming Places*. (London: Routledge).
Walsh, K. (1992), *The Representation of the Past: Museums and Heritage in the Post-modern World*. (London: Routledge).
Webb, J. (2009), *Understanding Representation*. (London: Sage Publications).

PART I
Relocating the Visual

Chapter 2
Inside/Outside: Ways of Seeing the World

Tony Schirato and Jen Webb

Contemporary culture is, media commentators, educators and cultural theorists inform us, visual culture; and this brings to the fore a particular problem of knowledge: how do we see and know what it is we see? According to the art historian Jonathan Crary, contemporary visuality is undergoing a transformation '... more profound than the break that separates medieval imagery from Renaissance perspective' (Crary 1990, 3). Crary argues that this process must be understood as an imbrication of technologies, techniques, subject positions and discursive orders; and he provides a comprehensive list of the specific developments he sees as simultaneously constitutive and characteristic of this new visual regime:

> The rapid development in little more than a decade of a vast array of computer graphics techniques is part of a sweeping reconfiguration of relations between an observing subject and modes of representation that effectively nullifies most of the culturally established meanings of the terms observer and representation. The formalization and diffusion of computer-generated imagery heralds the ubiquitous implantation of fabricated visual 'spaces' radically different from the mimetic capacities of film, photography and television. These ... corresponded to the optical wavelengths of the spectrum and to a point of view, static or mobile, located in real space. Computer-aided design, synthetic holography, flight simulators, computer animation, robotic image recognition, ray tracing, texture mapping, motion control, virtual environment helmets, magnetic resonance imaging and multispectral sensors are only a few of the techniques that are locating vision to a plane severed from a human observer. Obviously other older and more familiar modes of 'seeing' will persist and coexist uneasily alongside these new forms. But increasingly these emergent technologies of image production are becoming the dominant modes of visualization according to which primary social processes and institutions function (1990, 1–2).

This technological shift is thus associated with an abstraction of vision from human beings and its relocation to the technological plane. This is not to suggest that vision has ever been unmediated; people's capacity to see and perceive has always been framed by norms, conventions and rules, and dependent on cultural contexts and on training. But something more is at stake now: the imbrication of new technologies, techniques and discursive orders has brought about not just a 'problem of vision' (Crary 1990, 3) but also the problem of the observer. With the abstraction of vision from the body, it may be that the body itself is losing its integrity in ways not previously considered. With the combined subjectification of vision, and problematizing of the viewer, it is possible that the subjects themselves could disappear behind the networks of digital information.

In his analysis, Crary is not simply suggesting that technical and technological factors (the move from analogue to digital culture, the dislocation of space) are agents of change. Rather, he is at pains to demonstrate how technical changes and technological developments (from the seventeenth century onward) are: first, part of the programs, values, incitements and discourses associated with modernism and modernization; second, commensurate with and derived from modernism's most prominent and influential visual regimes (normalization and discipline; capitalism; science); and third, connected to the promotion of the visible as the main criterion for negotiating questions and problems of truth, authenticity, proof and value.

Crary uses Foucault's notion of the modern discursive regime, which was predicated on the development of scientific categories, knowledge, techniques and attitudes, in order to suggest how and why new ways of understanding vision came about at the same time. The development of scientific techniques, procedures, logics and forms of knowledge not only increased what could be seen, it also changed the way the practices and mechanisms of seeing were understood. Prior to the modern period, perceptual experience was considered to be largely 'given' to us by and from the external world; in other words, we more or less received the truth of the world, rather than saw it in a subjective way. Science in the modern period, however, produced a different way of understanding perception, what Crary refers to as 'the idea of subjective vision', which involved 'a severing (or liberation) of perceptual experience from a necessary relation to the external world'. As Crary writes:

> … the rapid accumulation of knowledge about the workings of a fully embodied observer disclosed possible ways that vision was open to procedures of normalization … Once the empirical truth of vision was determined to lie in the body, vision (and similarly the other senses) could be annexed and controlled by external techniques of manipulation and stimulation (1990, 12).

The insight here is both quite profound and paradoxical: the realization that the basis of vision lay 'in the body' and was thus subjective arose at the same time as the development of science and its objective and objectifying ways of seeing, categorising and normalizing people and populations.

This chapter will follow through and explain some of the main ramifications of this insight, in particular the cultural technologies of seeing that inform understandings of visual culture, and will explore issues relevant to the context of heritage-based research. The basis of the account, exemplifications and discussions that follow is Bourdieu's observation that while 'there is no way out of the game of culture' (1989, 12), some level of objectification is possible if one is willing and able to interrogate and understand the relation between what is seen and the authorized ways of seeing (see Waterton this volume).

Analysing Vision: The Linguistic Turn

Central to the work of normalizing vision and establishing conventions and techniques for seeing are the analytical frameworks through which visual material can be understood. This shifts the discussion away from the physical realm of eyesight and neurological perception to the modes by which we make sense of what we see. Academics give names to these modes, and lay out the various steps and processes applied; most people in their everyday lives do not name these modes or apply them consciously but, nonetheless, through the processes of normalization of thought, they are likely to make sense of what they see in a limited number of ways. One of these is semiotic analysis. Although we have made the point above that the world is now increasingly one of visual rather than literary culture, this does not necessarily mean that we know what we are doing when it comes to the work of perception, particularly since seeing is a social activity, predicated as it is on the dispositions to see and make sense that are part of the habitus of each individual in a culture. Therefore, seeing is also always a changing activity, subject to the dynamic forces of culture and society, and its changing perspectives on what is important, what is visible, and what should be understood from the world around us.

If seeing is as much a social as a natural process, then how we see must be managed – like all social practices – both through self-regulation applied by individuals, and through the various mechanisms of control that, as Foucault has shown in his works on power and disciplinary practices, thread through society. One of the ways of managing the process of seeing is by determining what visual texts might mean, and how they can be read, and here we turn to the technologies of 'reading' the visual. Martin Jay (1993) points out that though we are living in a period that is filled with visual texts, we are in fact living in a deeply non-visual period because we make sense of the world by using non-visual analytic devices. This is certainly evident if we look to twentieth century scholarship, which is marked by what is called the 'linguistic turn' – a move within the humanities to use the analytical modes associated with literary texts to make sense of society, visual images, individual psychology, and so on. Under this approach, all social practices were understood as meaning-making practices, or semiotic events (Evans and Hall 1992), and visual texts were considered to communicate according to linguistic rather than iconographical rules. Certainly semiotics is a useful analytical technique: by reducing everything to signs, it draws attention to the property of readability in all that surrounds us, and to the fact that we do not simply see, but actually make sense of, what is in our visual sphere. It also draws attention to the principles of difference in all human communication or meaning making; principles many people, in fact, apply in a naïve or unconscious manner when looking at the world around them.

Figure 2.1 Streetscape: Wanchai, Hong Kong, 2006

In viewing Figure 2.1, a streetscape in Wanchai, Hong Kong, for instance, we can read it as 'sign' by focusing on the principles of difference. An important starting point is to differentiate the text from other texts in order to secure its meaning. So, we can confirm: it is not a landscape, it is not a family portrait, it is what it is – a streetscape – in no small part because it is not something else. Difference operates also to make sense of the signs, or objects, that are juxtaposed within this text: note the hard vertical lines of the buildings set against the soft, fernlike palm trees planted in the median strip, or the matt, absorbent quality of those fronds compared with the glossy surfaces of the buildings. The former allow connection and contact; the latter speak not just of mirroring, but also of impermeability: they allow no access, not even for air. One reading of these sets of differences is of the triumph of capitalism and technological development over nature, in the dominance of roads and buildings over the fragile trees – the only signs of nature in all that space. Certainly a semiotic analysis might suggest this text as a celebration of capitalism.

But it is also possible to read the image not in terms of the difference of semiotics, but as a story shaped by analogy, and to see the streetscape as part of a totality that makes up the image of an urban space. The multi-lane roads and cars go together to tell a story of the movement of many people around the city spaces; the tall glass-and-steel buildings are statements not just of modernity and capitalism, but also of accommodating cultural activity within nature's constraints: squeezing floor space into a limited urban area. The palm trees in the median strip support this story of the pressures of nature on culture – and vice versa – and of their imbrication within one another. The (comparative) height of the palm trees and the vertical quality of their trunks are mirrored in the height and verticality of the buildings, while the trees' lower level is matched and continued by the buildings on the horizon line.

We can read this image, then, in terms of semiotics and/or analogy, and make sense of the image not only by evaluating difference – the digital on-off/same-different of the linguistic model – but also through the combination of internal elements. It is necessary, indeed, to look beyond semiotics in reading visual texts. Semiotics is, after all, a linguistic model and, as Roland Barthes (1977) points out, the visual may be like a language, but it does not possess the grammatical elements, organization and structure that we expect from words. As such, the work of reading visual matter is not simply a matter of observation, but a constant and constantly adjusted way of fitting the material world to whatever constitutes the 'truth' of society as it obtains in a particular context, under a particular visual regime, and a constant and constantly edited narrativizing of that visual material.

Analysing Vision: The Narrative Turn

Narrative logic offers a fresh way of reading a visual artefact though, as with semiotics and analogy, possible readings depend on the content of the text, who reads it and where it is read; in short, its cultural context. Each reader of the visual text-as-narrative will bring with them their own interests, concerns and histories, and also the way in which they have been 'produced' by discourses of governmentality and normalization, to make sense of what they see. The image

Figure 2.2 Jonathan Borofsky, *Man Walking to the Sky*, 1990, Kassel, Germany
Source: With permission of Jonathan Borofsky

of Borofsky's *Man Walking to the Sky* (Figure 2.2) is obviously a piece of visual culture; it is a work of art produced by a recognized artist, selected for exhibition in Documenta (one of the more important international art shows), and purchased by the city of Kassel, where it has the status of a 'must see' cultural icon. But this does not complete its story; various commentators offer their own readings of it. Gernot Böhme, for instance, writes about it from an architect's point of view: 'Borofsky's *Man Walking to the Sky* is simply an explicit rendition of what lines, beams, ledges, or ridge turrets do to space: they furnish it with a suggestion of movement' (Böhme 2005, 403). So for him, it tells a story of spatiality, and its elements focus on architectural technique. He is reading it, in short, from within the delimited field of his own discipline.

The artist, by contrast, reads it according to a very different narrative, as is clear from the following artist statement:

> This 80 foot steel and fiberglass sculpture was first shown at Documenta IX, in Kassel, Germany, 1990. But, the idea for this sculpture, no doubt reaches back into my childhood. When I was six years old, I used to sit on my father's knee, and he would tell me stories about a friendly giant who lived in the sky. The

> important thing about this giant was that he did good things for people. In his stories, my father and I used to go up to the sky and visit with this friendly giant every day (Borofsky 2004).

For him, then, it doubles as personal memoir and fairytale. The vantage point captured by Figure 2.2 certainly supports the latter, with the sheep looking curiously across at the Man.[1] It is thus a story when read by an artist, and an exemplification of lines in space when read by an architect. One focuses on the man/giant, and the other on the projection of steel into space. Each reading is valid, in its own terms; but neither is true in the formal sense of the term. Rather, each responds to the narrative inflection that is necessarily part of the object and the image it conveys.

Note, here, the qualification embedded in the phrase 'narrative inflection' rather than being a narrative itself. It does not constitute a story in terms that would be acknowledged by narratologists, for whom a text is 'narrative' when it not only tells someone about something or someone, but is structured to tell in a relatively coherent fashion, including temporal and causal elements. *Man Walking to the Sky* is a text with narrative potential, but is not a story per se because it is able only to provide springboards for stories that the readers/viewers must produce for themselves.

Among these tools for making and reading story are plot, character, time, event and causality. The representation of a human figure is a good way of conveying character in visual texts, and conveying action and mood. The position of the body and limbs, and expressive movements such a smile, a look of fear or sorrow, a hand extended in friendship or a fist raised in anger, establish an obvious appeal to empathy in images, as Gombrich explains (1982, 84–6). Narrative can also be implied or identified in visual material by devices such as the arrangement of the iconography or the use of perspective to provide a central focus. Lines, shapes and angles help to tell the story; and the use of light, particularly, structures the reading of the narrative. Lighting draws attention to particular features in a text and ensures we make sense of the images. Bright colours and a whimsical style, for instance, create a light, possibly fantastical sense; dark images convey melancholy or threat; while black-and-white images immediately signal a particular aesthetic.

All are obvious in *Man*. Its imaginary, indeed impossible, quality means that no more information can be made available to us than is contained in the work itself and the artist statement. But, as we have pointed out, that statement is no more necessarily 'true' than any other commentary. This means we can only guess at the story on the basis of its narrative clues. Character is evident in the presence of a man, casually but smartly dressed, marching confidently up the pole towards the sky. We can read him as fitting into a number of categories: middle class, young to middle-aged, white, male, in control of his own destiny. The steepness of the angle of his ascent suggests effort, and the sheer height he reaches has a triumphalist

---

1 The sheep are part of a memorial to the animals killed during Second World War air bombardments in Kassel.

sense, but the ordinariness of the man himself provides a point of connection with 'the Everyman', and thus a sense that we too might take this journey. The man himself, and his pole, are relatively light in colour, pointing to the 'good giant' motif presented by the artist, and also helping to keep the mood light and breezy. Finally, the sense of movement in a work implies event, and as such, it also implies time and causally related events: where is the man coming from and what is he leaving behind? What was the impetus for him to take off like this, and where is it likely to end? What happened, or what is going to happen? In short, this visual artefact has elements of story; it points us in a particular direction, or series of directions, but it can't 'tell' in the way a conventional narrative would. To fit the category of 'narrative' it would have to be crafted and comply with particular generic formulae, patterns or design 'tools'.

The two related principles of genre and intertextuality go a long way towards framing the meaning of a visual text because images, like any texts, do not constitute a pure field. That is, we don't come to them innocently, but always read them with reference to all the other things we know and have seen, and with which we are familiar. This means that no text is entirely free-floating, or entirely subject to the whim and imagination of its viewer to make meanings, or tell stories; it will always place constraints on its viewers. The *Man*, for instance, could be seen as part of a science fiction narrative (he is, after all, walking to the sky), but he carries no scientific or technological paraphernalia, which puts a damper on that reading. The action of walking up a steep pole might suggest an action story, but we would be hard pressed to say that the man is fleeing a disaster or engaged in drama: the calm expression of his face and body, and his confident open stance, preclude that reading. We are left with an image that combines the quotidian with the fabulous. Perhaps, then, we can add to the list of possible readings of this work that it is a story about human potential, about freedom, about the marvel that is everyday life. In any event, we read this narrative both in terms both of the content of the work itself, and by drawing on principles of intertextuality and genre. In other words, we make sense of it on the basis of other texts with which we are familiar. And because no social practice can operate in isolation from its social context, any spoken, written or visual text will either connote or cite other texts, and by recalling these known stories, they will propel our reading in a particular direction. This, in itself, is part of the normalizing work applied to ways of seeing the world.

Science and Seeing

Central to contemporary approaches to visual perception is a phenomenological awareness, or recognition of how embedded the physical and experiential is in any reading of visual material. We 'know' what we see, Bourdieu suggests, largely because we are at home in the world, and at home with seeing:

> The agent engaged in practice knows the world ... too well, without objectifying distance, takes it for granted, precisely because he is caught up in it, bound up with it; he inhabits it like a garment ... he feels at home in the world because the world is also in him, in the form of the habitus (Bourdieu 2000, 142–3).

Certainly a great deal of perception works, as Bourdieu argues, at the level of the body; seeing is concerned as much with feelings as it is with sentences and stories, and it involves our whole being, not just our abstract intellectual identity. This approach might counter the abstraction of visuality described by Crary, and restore the human and the human eye into the work of seeing because, after all, seeing/reading the visual is, as Maurice Merleau-Ponty writes, a very physical activity (Merleau-Ponty 1962, 407), and reading visual material involves the body and the emotions. It is a sensate rather than a purely intellectual or abstract means of communication, one that involves the person, their own context, tastes and interests, as well as the visual regime in which they have been constituted.

Despite this, the conventional perspective on vision is still more attached to the scientific model than to phenomenology. The qualities that Martin Jay identifies as leading to the supposed ocular-centricism of Greek culture – a sense of detachment and objectivity, apprehension without the need for proximity, and a 'prospective capacity for foreknowledge' (1993, 25) – were both the basis of, and further developed by, Western science, and in many ways they still direct our understandings of visual perception. Contemporary ocular-centricism was extrapolated from sixteenth- and seventeenth-century scientific principles championed by practitioners such as Francis Bacon and Robert Boyle, who emphasized techniques predicated on the observable replication of phenomena in isolated, and therefore uncorrupted, contexts such as the laboratory. The meaning of any visual material is thus extracted from its context and, indeed, from its own particularities, and filtered through the techniques and norms of observation that have constituted the observer.

This is very different from pre-modern modes of perception. Up to the seventeenth century, Western culture understood the world as a book, or something to be read, and the obverse is true too: just as the world is a book, so too a book can be the world – especially the book of law, or the Bible. In that period, 'the prose of the world' operated according to notions of resemblance and similitude (Foucault 1970, 19). Everything, that is, could be read by the recognition and application of a set of categories that existed not in the utopia of contemporary representation, but in the natural order, where similarities established relationships and generated knowledge and understandings. But this regime increasingly comes to be seen as unreliable, and by the seventeenth century a new perspective on visual perception was starting to take hold. For Foucault, the adventures of Don Quixote (in the novel by Miguel de Cervantes, first published in 1604):

> ... mark the end of the old interplay between resemblance and signs and contain the beginnings of new relations. Don Quixote is ... the hero of the Same. He

> never manages to escape from the familiar plain stretching out on all sides of the Analogue (1970, 51).

Quixote is the 'hero of the Same' because he is entirely immersed in similitude, to the point that he is unable to distinguish objects from words; reality lies, for him, in the romances of the books he reads, and behind all he sees and does is the attempt:

> … to transform reality into a sign. Into a sign that the signs of language are really in conformity with things themselves. Don Quixote reads the world in order to prove his books. And the only proofs he gives himself are the glittering reflections of resemblances. His whole journey is a quest for similitudes: the slightest analogies are pressed into service as dormant signs that must be reawakened and made to speak once more. Flocks, serving girls and inns become once more the language of books to the imperceptible degree to which they resemble castles, ladies and armies (1970, 52).

When Quixote is confronted by the unreality of these resemblances, he resorts to the notion of magic as an explanation: it is sorcery that breaks the logic of similitude, that deceives us into thinking we see difference instead of the same.

Early modern science offers a completely different explanation. By the seventeenth century, the logic of resemblance had come to seem a space for confusion and enchantment. For Bacon, resemblance is a kind of fakery, an illusion, an idol. 'The idols of the den and the idols of the theatre,' Foucault explains, discussing Bacon's point, 'make us believe that things resemble what we have learned and the theories we have formed for ourselves; other idols make us believe that things are linked by resemblances between themselves': they are, in short, 'spontaneous fictions of the mind' (1970, 57), and here we have an early rejection of the 'seeing is believing' dogma. In the writings of Bacon and his contemporaries, then, is the start of an attachment to empirical proof and scientific rationalism. From the seventeenth century onward, science will only see through and believe in analysis, evidence and trained perception. But it is also, as Jean Baudrillard (1993, 51) points out, the start of what he designates 'forgery … from the deceptive finery on people's back to the prosthetic fork, from the stucco interiors to Baroque theatrical scenery. The entire classical era was the age of the theatre par excellence'. So, smuggled in at the birth of scientific reason is the disconnection of seeing from what is there. The disenchantment of the world that is an effect of the scientific gaze is accompanied by the 'disenchanted universe of the signified' (Baudrillard 1993, 50).

Seeing under Capitalism

For Baudrillard, the traditional, pre-modern and pre-capitalist discourses and perspectives that categorized and evaluated the world were limited, and their meanings were bound by such factors as membership of classes such as the aristocracy or the clergy, or the possession of cultural capital such as literacy or a university education. This meant that signs could not be arbitrary, but were bound by social obligations and relations. This is no longer the case. Instead, signs are arbitrary, bound not by context or relationship but by the logic of competition, and of simulacra. Disconnected from resemblance, let loose from empirical connection, they now merely convey counterfeit[2] – the game of connection and of simulation.

The simulacrum is, for Baudrillard, the complete disentangling of signs from the material world; it is:

> ... the generation by models of a real without origin or reality: a hyperreal. The territory no longer precedes the map, nor survives it. Henceforth, it is the map that precedes the territory – precession of simulacra – it is the map that engenders the territory (Baudrillard 1994, 1).

The image of Wanchai (Figure 2.1) shows something of this effect. The buildings are, as we noted earlier, entirely mirrored. They respond not to the environment, but to constantly bounced-back images of each other, as is evident in the tall building on the left. As signs, they are more real than reality; indeed, it is difficult for reality to get a word in edgeways, in this space where there is little but an endless reflection of reflections. Added to the counterfeit nature and hyperreality of the site, the buildings are overlaid with additional signs. On the right, for instance, it is possible to see the outline of a drawing in neon that flashes in front of the mirrored wall, adding to the cacophony of reflections, while below that is a corporate logo, signalling the real work of these buildings and of their surfaces: to speak not of actuality, of the steel and concrete that is their material, but of the flickering, counterfeit nature of their business – a simulacrum of financial exchange.

Contrast this with the image of (a tiny portion of) Salisbury Cathedral (Figure 2.3). This was conceived and produced under a very different economy – pre-democratic, pre-capitalist – and under the logic of distinctions and values such as high and low culture, originality, authority and taste. These values and hierarchies constituted the grid through which people saw the world and their place within it. In that (Medieval) visual regime, the dominant classes produced a very limited and relatively unambiguous series of signs and texts that represented, articulated and authorized social relations, duties, and modes of behaviour. This is illustrated particularly well by Medieval English or French churches and castles like Salisbury Cathedral (completed and dedicated in 1258). It is almost impossible to escape the visual exchange such texts initiate; one looks at them from a position of relative insignificance (the viewer must, in every sense, look up to them), while they look

2 It is perhaps worth pointing out the consistency with which we see, in the literature, an anxiety over the truth of the sign: from Don Quixote's logic of the sorceror, through Bacon's concern about idols, to Baudrillard and the forgery, it seems we are always looking for 'the real' in what we see.

Figure 2.3 Section of Salisbury Cathedral, 2005

back down from their place of grandeur, power and permanence. They are the sign of a power relationship that translates into every other part and activity of the social sphere; few could doubt, or speak against, the power manifested in these buildings, or the soaring narrative they imply.

They are also embedded in the actuality of their environment. Made of local stone, their colour blends into, rather than bounces back, their surroundings, emerging as it were from the landforms in which they are located. They are, in many ways, the world as book, as well as being effectively and politically the book that ruled the known world. Such a building, read and experienced as visual text, was unambiguously a sign of God's greatness and power, which was passed along the chain of signs associated with the cathedral – the clergy, their costumes, the texts they produced, the ceremonies they presided over and the words they spoke. While there are many imposing modern buildings, they operate under a different – a democratic, a market driven – logic. They do not resemble the grandeur of divine authority, as does a cathedral. Rather, they simulate spectacle. The sight of the skyscraper canyons that dominate modern cities, as seen in the Hong Kong streetscape, is just as likely to elicit a sense of disenchantment as it is wonder or awe.

For Baudrillard, the most important issue is that we see, and are seen, through a visual regime where signs are little more than exchange objects, infinitely equivalent to other signs. The notion of the authorized, and even transcendent, visual regime, which organized and guaranteed value and distinction in the Medieval period, is replaced by what Baudrillard calls 'the emancipated sign', or the promiscuous mixing and 'proliferation of signs according to demand' (1993, 51).

The notion of capitalism as a visual regime is probably most famously theorized and articulated in the work of the Frankfurt School's Theodor Adorno and his contemporary Walter Benjamin. Benjamin writes in his famous essay, 'The Work of Art in the Age of Mechanical Reproduction', that the capitalist system and its technologies (particularly mass production) strip away the aura that provides unique cultural artefacts with their supposed innate value. He saw this process as a potentially liberating development, because it 'detaches the reproduced object from the domain of tradition' (Benjamin 1968, 215), undermining notions such as originality and genius, and thus facilitating a break from the authority of high cultural institutions that served as the arbiters of cultural value. In other words, a whole order of things and meanings that were dependent on the authority of exclusivity were replaced by the democratizing free flow of the market. Texts that were authorized as valuable, such as paintings or sacred texts, and were the possessions of, and thus associated with, elite groups, were now free to circulate throughout society as copies. Everyone could now own a Rembrandt or Vermeer print, or a replica of the Venus de Milo. But what was more important was the effect this change would have on the way people saw the world: works of art or sacred texts were no longer part of a chain of signs that inevitably led to, and authorized, the worldviews of the elites.

For Theodor Adorno (1981), these processes, and in general the market interventions in the field of culture that have turned art works into commodities, brought about the denigration of human faculties and virtues, and undermined the capacity of visual culture to break with the normalizing and controlling discourses of government and the market. He writes:

> [T]he culture industry intentionally integrates its consumers from above. To the detriment of both it forces together the spheres of high and low art, separated for thousands of years. The seriousness of high art is destroyed in speculation about its efficacy; the seriousness of the lower perishes with the civilizational constraints imposed on the rebellious resistance inherent within it as long as social control was not yet total (1981, 98–9).

Adorno looks back to the pre-capitalist period, when 'high art' objects – Bosch's paintings, say, or Michelangelo's sculptures – were recognized and pronounced as valuable by cultural elites, and served as exemplars for the wider population. Of course, another of their functions was to reproduce and naturalize regimes of power and value; think, for instance, of the manner in which Italian Renaissance art helped authorize and legitimize the power of the Catholic Church. But Adorno argued that certain works of art (usually those that have survived across history) draw attention to the limits that a society places on what and how things are seen. That is, they both reproduce the worldviews that they are produced by and within, and go beyond them, providing viewers with a way of seeing outside the constraints of contemporary discourses of normalization and control. However, once questions of cultural value are taken away from authorized institutions and practitioners, and moved into the domain of the marketplace, he argued, people would only be exposed to texts aimed at the lowest common denominator, texts that Adorno saw as simplistic and sensationalist, and which would not challenge them to think beyond a normalized worldview. This is a view that still resonates with many art commentators: Robert Hughes, for instance, recently denounced artist Damien Hirst's 'simple-minded and sensationalist' works (2008a), insisting that: 'art should make us see more clearly and more intelligently … if art can't tell us more about the world we live in, then I can't see any point in having it' (2008b). While this sparked a round of media-based squabbling about the value of art, it does make it clear that visual media still matters, and that the question of how to read and evaluate the visual remains unresolved.

Consuming Spectacles

It also, of course, demonstrates the importance of spectacle in practices associated with the visual domain. We suggested, at the beginning of this chapter, that modernization, its visual regimes and, in particular, its incitement to the visual has wider applicability than just the domain of mere technological change. Instead, modernization and its effects are germane to the development and character of the wider sociocultural field. Theorists such as Jonathan Crary (1990, 1999), Paul Virilio (1994) and Martin Jay (1993) have demonstrated that the relationship Foucault identifies between discipline/normalization and forms of knowledge was accompanied by a separation of sight and seeing from its material contexts, and a quantification, standardization and abstraction of the observable world. As

Crary writes, this 'autonomization of sight, occurring in many different domains, was a historical condition for the rebuilding of an observer fitted for the tasks of "spectacular" consumption' (Crary 1990, 19).

This reference to spectacular consumption constitutes a borrowing from Guy Debord's notion of the society of the spectacle (Debord 1977, 42). Spectacle is, in Debord's terms, the replacement performance, in and through the mass media, of the sociocultural field and its ideas, values, worldviews and forms of subjectivity. It is 'the moment when the commodity has attained the total occupation of social life. The relation to the commodity is not only visible, but one no longer sees anything but it: the world one sees is its world' (Debord 1977, 42). It is nothing less than (predominantly visual) technology employed simultaneously to produce and occlude an arrangement and architecture with regard to all visual practices and dispositions.

There are four main interrelated conditions or characteristics of the spectacle. First, attention is all important – it must be attracted and maintained. Second, vision is arranged, organized and disposed within various hegemonic visual regimes, the most influential and pervasive of which is that of capitalism. Third, everything is (potentially) reduced to the status of commodity, and there is an emphasis on necessary, repetitive and mobile (visual) consumption. Fourth, the subject-as-spectator relates to the social and the self through the consumption of commodities. This process stands in for, and functions as a simulation of, the social. In other words:

> Spectacle is not primarily concerned with a looking at images but rather with the construction of conditions that individuate, immobilize, and separate subjects, even within a world in which mobility and circulation are ubiquitous. In this way attention becomes key to the operation of noncoercive forms of power (Crary 1999, 74).

Debord's notion of the spectacle can be understood as a version of Foucault's theories of the relation between new visual techniques and regimes and processes of surveillance, discipline and normalization. Today, when we look at and evaluate people, objects and practices, we usually do so using these and other categories, discourses and logics related to, or taken directly from, the human sciences. And this tendency is accentuated as these scientific categories of truth and reality are circulated throughout popular culture, particularly in and through the media-as-spectacle.

The reproduction and continuation of normalized ways of seeing is in a sense a circular process: social fields, institutions, techniques and mechanisms produce subjects who are inclined to see and understand the world, and everyone in it, in terms of recognizable and authorized categories, and the templates that go with them. The American gender theorist Judith Butler (1993) refers to authorized, iterative performances of normalization. Butler's argument is that there are sites in a culture (most particularly the media) where we can check out that our subjectivity is 'on track' (that is, does it remain true to itself while taking into account changes in society?) in terms of our body shape, clothes, mannerisms, or ways of seeing and evaluating other people. These images, ideas and performances constitute a vast store of up-to-date templates for, or models of, a normal, healthy, attractive and desirable subject. This evaluation and categorization of every subject, including the self, is 'neither a single act nor a causal process initiated by the subject … Construction not only takes place in time, but is itself a temporal process which operates through the reiteration of norms' (1993, 10).

Cultural texts are filled with types that let us know what is attractive and desirable, healthy and normal. But they do more than offer us role-model images of bodies, homes and possessions, or of culturally significant artefacts and sites – they help us understand, negotiate and see these places, persons and objects. We see images performing this role in magazines and on television, in newspaper articles and scientific or government reports, and in brochures, marketing information, websites and other publications associated with the cultural heritage industry. Such images, along with the texts that support and explain them, offer representations of icons that do more than simply picture them: they also inflect and interpret them, and thus instruct viewers in how to read them.

Consider, for example, the flyer for the 2008–2009 British Museum exhibition, *Babylon: Myth and Reality*.[3] The title, layout of the flyer, quality of the card, selection of image and colour used for the text box all combine to produce not only marketing, not only an aesthetic object, but also a normalizing and authorized response to the work. Myth and Reality: these related terms inform us that the exhibition comes complete with the entrancing qualities of story and the approved elements of fact. The layout of the flyer – in landscape format – encourages a slow reading as the eyes are drawn from left to right across the surface. Attention is focused on the image that covers most of the flyer, from the left margin to the text box at the far right: a mosaic of a snarling lion. In tiny font at the bottom right, a caption informs the reader that this is a 'glazed brick relief of a lion from Babylon's Processional Way', but even without the text, those with the most basic knowledge of Babylon or the Mesopotamian region would know they were viewing an object of value and antiquity. Its value is signalled by the selection of this image to take up most of the front face of the flyer, and by the quality of the card used for the flyer and its finish; its antiquity is signalled by the combination of 'Babylon' and 'British Museum' – both part of heritage rather than the contemporary world – and by the style of the lion and its mosaic form. The turquoise that still colours many of the bricks has been extended into the text box of the right of the flyer, which simultaneously associates the British Museum with the majesty of ancient Babylon, and affirms the provenance of the mosaic (the Museum being an authority on works of antiquity, it thus assures us that we are seeing an important original). On the back of the flyer are listed related events, both instructional and entertaining,

3 This is available as hardcopy, and also on the Museum's website at http://www.britishmuseum.org/pdf/BABYLON_Family_perf_lates_films.pdf.

along with more information about the exhibition and the British Museum itself, and this is tied to the important front page by the colour and typeface.

What this document suggests throughout is the authority invested by the Museum in the exhibition and associated materials (lectures, books, films), and the unquestioned significance of the works on display. No overt material points to the politics of displaying an exhibition on Babylon and signifying it through the image of a snarling lion, during the period when Britain is at war or in a state of political tension with nations in the region of Mesopotamia. Nothing is made of the fact that a document so engaged with commercial activities (including the costs of attendance and participation in various of the activities) and institutional imperatives (points of entry, memberships, opening hours) seems to rise above the tawdry concerns of the everyday world, promising viewers a taste of the glories of an ancient past.

It is easy to deconstruct such a text; but the impact of the splendid lion whose image absorbs most of the front page, and hence most of the viewer's attention, and the storied quality of the name of Babylon, together encourage a particular viewing of the text. This is one that accepts without question the value of antiquity and its exhibition in the present day; that accepts without question the authority of the British Museum; that is disposed to accept the regulations associated with visiting the Museum and the perspective on Babylon that is offered by its curators and designers. This is not to criticize the Museum or its flyer, but rather to make the point that we know the world and what we see in it, not first because of our direct experience, but through the representations made of it, especially by authoritative institutions. Writing on cultural tourism, John Taylor noted that '[p]hotography and language is conventionally used to keep a dream of England before waking eyes' (1994, 4). So too, photography and language are conventionally used to keep before our eyes a dream – the privileged, authorized and/or preferred perspective – of culture, of heritage and of what constitutes value and distinction in that field. Over time, and through a variety of media, the relationship between the (normalized) ways of seeing and the audience as viewers is played out in this and similar ways: it returns, again and again, in the way the normal/normalized way of seeing the world is reproduced.

Let us leave the last word to Blaise Pascal, who in the seventeenth century prefigured much of what we have suggested here about the distance between observer and observed, and the limits on our capacity to see and make sense of anything in an objective manner:

> For we must not misunderstand ourselves: we are as much automaton as mind. And therefore the way we are persuaded is not simply by demonstration. How few things can be demonstrated! Proofs only convince the mind; custom provides the strongest and most firmly held proofs: it inclines the automaton, which drags the mind unconsciously along with it. Who has proved that tomorrow will dawn, and that we will die? And what is more widely believed? So it is custom which persuades us ... In the end, we have to resort to custom once the mind has

seen where the truth lies, to immerse and ingrain ourselves in this belief, which constantly eludes us. For to have the proofs always before us is too much trouble. We must acquire an easier belief, one of habit, which without violence, art, or argument makes us believe something and inclines our faculties to this belief so that our soul falls naturally into it (Pascal 1995, 661).

References

Adorno, T. (1981), *The Culture Industry: Selected Essays on Mass Culture*, edited and introduced by J.M. Bernstein (London and New York: Routledge).

Barthes, R. (1977), *Image Music Text*, trans. S. Heath (New York: Hill and Wang).

Baudrillard, J. (1993), *Symbolic Exchange and Death*, trans. I.H. Grant (London: Sage).

Baudrillard, J. (1994), *Simulacra and Simulation*, trans. S.F. Glaser (Ann Arbor: University of Michigan Press).

Benjamin, W. (1968), *Illuminations*, edited and introduced by H. Arendt; trans. Harry Zorn (London: Pimlico).

Böhme, G. (2005), 'Atmosphere as the Subject Matter of Architecture', in Philip Ursprung (ed.) *Herzog and De Meuron: Natural History*, (New York: Springer) 395–406.

Borofsky, J. (2004), 'Man Walking to the Sky', [Online]. Available at: http://www.borofsky.com/pastwork/public/manwalkingtothesky[germany]/index.html [accessed: 1 September 2008].

Bourdieu, P. (1989), *Distinction*, trans. R. Nice (London: Routledge).

Bourdieu, P. (2000), *Pascalian Meditations*, trans. R. Nice (Cambridge: Policy Press).

Butler, J. (1993), *Bodies that Matter*, (New York and Cambridge: Routledge).

Crary, J. (1990), *Techniques of the Observer: On Vision and Modernity in the Nineteenth Century*, (Cambridge, MA: MIT Press).

Crary, J. (1999), *Suspensions of Perception: Attention, Spectacle and Modern Culture*, (Cambridge, MA: MIT Press).

Debord, G. (1977), *The Society of the Spectacle*, (Detroit: Black and Red).

Evans, J, and Hall, S. (eds) (1999), *Visual Culture: The Reader*, (London: Sage Publications).

Foucault, M. (1970), *The Order of Things: An Archaeology of the Human Sciences*, (London and New York: Routledge).

Gombrich, E.H. (1982), *The Image and the Eye: Further Studies in the Psychology of Pictorial Representation*, (London: Phaidon).

Hughes, R. (2008a), 'Day of the Dead', *The Guardian*, 13 September, 8.

Hughes, R. (2008b), 'The Mona Lisa Curse', *Art and Money* (series dir. Mandy Chang), Channel 4 Television, UK (viewed 21 September 2008).

Jay, M. (1993), *Downcast Eyes: On the Denigration of Vision in Twentieth-Century French Thought*, (Berkeley, Los Angeles and London: University of California Press).
Merleau-Ponty, M. (1962), *Phenomenology of Perception*, trans. C. Smith (London: Routledge and Kegan Paul).
Pascal, B. (1995), *Pensées and Other Writings*, trans. H. Levi, with introduction and notes by A. Levi (Oxford and New York: Oxford University Press).
Taylor, J. (1994), *A Dream of England: Landscape, Photography and the Tourists' Imagination*, (Manchester: Manchester University Press).
Virilio, P. (1994), *The Vision Machine*, (Bloomington and Indianapolis: Indiana University Press).

Chapter 3

People-Place-Past: The Visitor Experience of Cultural Heritage

Martin Selby

Despite the proliferation of research on cultural heritage tourism since the mid-1980s, there have been relatively few experiential studies. Within the literature with an experiential focus, common themes have included motivations and expectations (Poria et al. 2006), satisfaction (de Rojas et al. 2007), authenticity (McIntosh and Prentice 1999; Chhabra 2008), learning (Prentice et al. 1998), benefits (Beho and Prentice 1998), and commodification (Halewood and Hannam 2001). Recently, consistent with the social sciences in general, interest has shifted towards visitor encounters and the performativity of cultural heritage (Edensor 2000; Crouch 2000a; see Crouch this volume). It is argued in this chapter, however, that although such approaches have made important contributions to the conceptualization of cultural heritage experience, they remain rather tangential to understanding how, in a holistic sense, cultural heritage sites and landscapes are experienced.

Even within the more experiential 'non-representational' literature, something of a dichotomy exists between the visual and the performative. It is argued that this dichotomy – and others such as physical and psychological, being and appearance – constitutes a challenging dialectic in the context of cultural heritage experience. The chapter, therefore, begins with a review of salient contributions from cultural studies, before turning to the philosophy of human experience: phenomenology. In order to explore the import of phenomenology for heritage studies, the chapter will be structured into six component parts: visualizing, representing, performing, perceiving, knowing and acting.

Visualizing

It is encouraging that in recent years human geographers have embraced the complex and embodied nature of vision (Bissell 2009; Cosgrove 2008; Doel and Clarke 2007; Wylie 2006). However, Urry's *Tourist Gaze* (1990; 2002) remains one of the most overt and influential attempts to conceptualize the visual nature of the visitor experience. Rather than repeating discussions of this much-cited work, it is more useful to consider some implications for conceptualising the experience of cultural heritage tourism. Drawing heavily on Foucault (1977), Urry's original contribution (1990) conceptualizes various socially-constructed gazes in relation to both touristic landscapes and the inhabitants of those landscapes. The Foucaudian influence is particularly apparent when Urry likens sightseeing to a form of surveillance. He even likens tourist destinations to eighteenth-century 'lunatic asylums' in England, where visitors could pay a penny to gawp at the 'mad behind the bars' (Urry 1990, 183). This aspect of cultural heritage experience is in evidence in cultural shows around the world. Simultaneously visual and performative, cultural shows purport to showcase local music, dance, customs, rituals and ceremonies to visitors. From Maori cultural shows in New Zealand, to Tyrolean evenings in Austria, visitors are entertained in a manner that exemplifies staged authenticity and commodification. Rather like the tourists in the film 'Cannibal Tours' (O'Rourke 1988), photographing and being photographed with the 'quaint', 'savage', or 'primitive' Other is a fundamental element of the experience.

Urry has often been accused of overstating the visual and developing a rather universal conceptualization of tourists (for example Jokinen and Veijola 1994). Other researchers have noted the tendency to undervalue the variety of tourist experiences in the pursuit of all-encompassing frameworks (see Morgan and Pritchard 1998; Selby 2004a), and the lack of ethnographic work underpinning typologies and classifications (Jamal and Hollinshead 2001). Despite an acknowledgement of these limitations implicit in the more performative and 'mobility' focused later edition (2002), the first edition of *The Tourist Gaze* was useful in linking representations of cultural heritage landscapes with the first-hand experiences of place consumers. Urry (1990, 2) urged researchers to evaluate 'the typical contents of the tourist gaze ...'. This focus on the signs and symbols of tourist landscapes and representations led to a longstanding interest in the semiotics of tourism, particularly in the context of cultural heritage.

The interest in semiotics – the science of signs and symbols – within the tourism and heritage literature is not surprising. Tourists seek out specific cultural markers, the signs and symbols that signify typical cultures and histories in a language familiar and desirable to different groups of place consumers. As Culler (1981, 127) noted 'all over the world the unsung army of semioticians, the tourists, are fanning out in search of the signs of frenchness, typical Italian behaviour, exemplary Oriental scenes ...'. Visitors not only seek out signs and symbols, but also capture them in photographs. Stallabrass (cited in Crang 1997, 361) noted how 'we might consider the 60 billion photographs taken each year as points of light on the globe, densely clustered around iconic sights and trailing off into a darkened periphery'. As Rojek (1997, 8) explains, the semiotic skill of the tourist involves 'the ability to move forwards and backwards between diverse texts, film, photographs, landscape, townscape and models, so as to "decode" information'. Visitors, however, do not merely consume the signs and symbols of cultural heritage tourism, but are themselves involved in acts of representation. Ringer (1998, 8) argues that the landscapes of tourism are 'articulated and made visible through the expression and acquisition of experiences'.

It is significant that semiotics developed from a simple 'landscape as text' model (for example Sauer 1925), to more socially-orientated genres that recognize that the interpretation of particular signs is culturally-specific (for example Gottdiener 1995). Contemporary approaches are more likely to recognize the non-material and symbolic aspects of culture, rather than merely the landscape. Authors such as Stuart Hall (1997) have encouraged researchers to analyse the codes by which meaning is constructed, conveyed and understood, and this is very salient to cultural heritage consumption. As Ringer (1998, 6) argues, the visual structure of the tourist landscapes 'expresses the emotional attachments held by both its residents and visitors, as well as the means by which it is imagined, produced, contested, and enforced'. Drawing upon pioneers such as Barthes (1984) semioticians are able to analyse the 'creation and reconstruction of geographic landscapes … through manipulations of history and culture' (Ringer 1998, 6). Shurmer-Smith and Hannam (1994, 13) go even further, suggesting that 'places do not exist in a sense other than culturally'.

An important distinction in the practice of semiotics is between the *signifier*, which is the expression carrying the message, and the *signified*, which is the concept that it represents. Thus, in the context of cultural heritage, a signifier might be an object such as an anchor outside a maritime museum; yet the signified may be 'ports, shipping and maritime history'. As Echnter (1999) explains, more socially-orientated genres of semiotics also introduce the 'interpretant', forming a semiotic triangle. Thus we can conceptualize another layer of meaning produced by place consumers, who may vary in the ways in which they interpret cultural heritage signs and symbols. A further aspect of semiotics (and linguistics) of significance to cultural heritage is the creation of a 'discourse' consisting of a consistently interpreted set of signs and symbols. As Morgan and Pritchard (1998) demonstrate, it is possible to identify a discourse that represents the people of Africa and South America as primitive, and the Welsh as antiquated and untalented. Crucially, a discourse relies upon both a consistent set of representations, and a 'textual community' that interprets the representations in a consistent manner. Not only is the discourse assumed to be natural and objective, but its reception is highly culturally and socially specific. This process of rendering innocent, or naturalizing, the representations and landscapes of cultural heritage tourism can be witnessed at cultural shows around the world, where staged authenticity is eagerly consumed by a compliant audience.

Representing

The demand for signifiers of particular histories and cultures on the part of cultural heritage consumers is met by the production of symbol-laden heritage landscapes. Destination marketing organizations, developers, heritage attractions, theme parks, resorts, and even shopping malls conspire to provide the signs and symbols so desired by place consumers. In the case of cultural heritage tourism, heritage sites are overt in signifying particular cultures, histories and social groups. In fact, the critique of heritage (see Hewison 1987) suggests that many heritage sites trade in signifiers produced by multi-media technology, rather than the physical artefacts characteristic of museums. This is apparent even in more traditional museums, which have increasingly adopted sophisticated methods of interpretation. The National Museum of Singapore, for example, relies heavily upon simulated (and dynamic) multi-media signs and symbols to signify the Chinese, Indian, Malay and European cultural mix that is synonymous with 'Singapore history'. Physical form, too, supplies place consumers with signifiers. It could be argued that the architecture of the Guggenheim Museum in Bilbao, Northern Spain, contains more salient, contrived, and powerful signifiers than the contents found inside. As the centrepiece of Bilbao's urban regeneration, it is renowned primarily for its iconic architecture. The same might be said of the Groningen Museum in the North of The Netherlands, which also stands as a signifier-laden statement of intent issued by the place marketers and developers of a peripheral city.

The deliberate manipulation of signs and symbols to produce cultural-historic landscapes for place consumers itself takes on various forms. Heritage sites and organizations promoting heritage often draw upon the 'phantasmagorical' in order to engage and entertain place consumers. As Pile (2005, 20) explains, phantasmagoria consisted of '...a procession of images that blended into each other, as if in a dream, and also concealed the means of its production'. Drawing heavily upon Benjamin (1939), Pile (2005) demonstrates the ways in which heritage managers, curators and place marketers create representations and landscapes that evoke dreams, magic (including voodoo), ghosts, and even vampires. Particularly powerful (and insidious) in Pile's analysis are examples from New Orleans, where the superficial and light-hearted 'voodoo tours' ignore and conceal the brutality of slavery. On a slightly different level, phantasmagoria has long constituted a staple within the promotional mix of stately homes and castles in the UK. It is difficult to cite a castle tour that does not include tales of torture, death, and, particularly, ghosts. As Selwyn (1996) demonstrates, myth-making is crucial to cultural heritage tourism. It is also an inherently representational and visual phenomenon.

Another visual manipulation of cultural heritage landscapes is concerned with creating hyper-reality – a contrived landscape that uses fantasy and imagination to construct a place that is actually more real than authentic places. Adding to the discussion of 'aura' provoked by Benjamin (2003), Rojek (1997, 59) points out the hyper-real representations created by cultural heritage can '...seem closer to everyday life than the original object'. Ritzer and Liska (1997) famously coined the phrase 'McDisneyization' to describe the search for the perfect simulation. In heritage sites, it is common to find simulated landscapes representing, for example, the Roman era, Greek history, or French lifestyle. A curious Malaysian example is at *Berjaya Hills*, where a French village called Colmar Tropicale 'will allow you to sample all things French, such as pizza and pastries from La Flamme Sidewalk Deli...' (Berjaya Hillls 2008). The Jorvik Viking Centre in York, Northern England, likewise offers '...a unique opportunity to ride in a time-capsule and take a journey through the streets of AD975 Jorvik' (York Archaeological Trust 2006). In constructing this hyper-real landscape, the creators feel it necessary to stress that 'everything here is based on fact, from the working craftsmen, the chattering noise of the gossiping neighbours to the smells of the cooking, the cesspit and the preserved 1,000-year-old Viking timber' (York Archaeological Trust 2006). The Beatles Story attraction in Liverpool, UK, is another attraction that creates a hyper-reality through assembling a plethora of auratic objects and Beatles signs and symbols in one place. In Gettysburg in the United States, Civil War re-enactments also aim to construct hyper-real landscapes. Despite their much-discussed staged-authenticity, for many consumers, cultural shows invoke an intensely hyper-real experience.

Another dimension of the cultural-heritage landscape is heteroptopia, defined as 'spaces that proliferate in a jumbled up manner on the same "level" as one another' (Philo 1992, 139). Influenced by Foucault's *Of Other Spaces* (1986), it would seem that many heritage sites create an experience that engages with a wide range of spatio-temporal sites and the relations between them. In heterotopia, actual times and places are 'simultaneously represented, contested, inverted' (Foucault 1986, 24), appearing to both abolish and preserve time. In heritage sites, this process is often facilitated by multi-media technology, whereby the visitor experiences a dizzying array of themes and periods of history. Although this can make for an entertaining experience, it has been dismissed as the creation of 'pseudo events' (Boorstin 1964), and 'a shallow veil that intervenes between our present lives and our history' (Hewison 1987, 135). The creation of heteroptopia, however, is another inherently visual practice.

Performing

In recent years, the social sciences have seen a non-representational shift, motivated by a critique of overly ocular representational studies. Non-representational studies are much more concerned with the body, other senses, and performances. Although this is often perceived as a shift away from understanding cultural heritage experience as a visual practice, it is argued in this chapter that visual and performative practices are simultaneous and complementary. Even researchers of the visual are increasingly seeking 'a more complex understanding of the visuality' (Degan et al. 2008, 2), acknowledging the embodied nature of visual practices. Not surprisingly, this has influenced how authors such as Crouch (2000a) and Edensor (2000) conceptualize the experience of cultural heritage, with much more emphasis on the embodied and performative nature of the experience. Bagnall's (2003) research at Wigan Pier, Northern England, for example, demonstrates the active nature of heritage consumption, and the ways in which visitors draw upon their memories and personal biographies to validate or contest the interpretation of exhibits. Authors such as Crouch (2000a) and Edensor (2000) have deliberately sought to reposition the analysis of cultural heritage towards sensuous (and sometimes sensual) experiences. Embodiment has been defined as 'a process of experiencing, making sense, knowing through practice as a sensual subject in the world' (Crouch 2000b, 68). Following a non-representational approach, therefore,

Figure 3.1 The View of the Tanneries in Fez, Morocco, without the Aroma

cultural heritage is grasped multi-sensually, with individuals expressing themselves though space, changing it in the process. This implies a more active involvement with place and space.

It is significant that even Urry (2002) develops the concept of 'sense-scapes' of tourist destinations. Urry (2002) is also concerned with a sixth-sense, 'Kinaesthetics', which informs the place consumer what their body is doing in space. Place consumers draw upon their imagination and memories to reconfigure heritage sites and landscapes. There is certainly evidence of this process at the National Museum of Singapore, where visitors on the 'personal history' route are encouraged through multimedia interpretation to reminisce and relate the exhibits to their memories and experiences. The various ethnic groups constituting Singapore's population create opportunities for this process, yet also create dilemmas in terms of what is left out or obscured. The embodied nature of cultural heritage experience is particularly striking in places such as Djemaa el Fna in Marrakesh. This large and busy square is a meeting place for a wide variety of inhabitants and visitors, including magicians, story-tellers, snake charmers, food sellers, dancing boys, fortune sellers, and medicine sellers. The experience of the square, and the adjacent media, is intensely multi-sensual (Figure 3.1). Also in Morocco, the tanneries in the medina of Fez make for a striking multi-sensual nature experience.

The view from the surrounding rooftops may be impressive, but a much more intense and lingering phenomenon is the smell of the hides and dyes, particularly the waist-high vat of white dye composed of pigeon faeces. The sprigs of mint handed out to tourists as they admire the 'view' add to the olfactory-ocular experience.

As Thrift (cited in Nash 2000) argues, a non-representational understanding recognizes the mundane practices that shape the conduct of human beings at particular sites. Such practices cannot be conveyed by texts, but are embodied in multi-sensual practices and experiences. There are certain places where practices are acted out, and these 'vortexes' (Thrift and Dewsbury 2000: 425) are often commodified as cultural heritage landscapes. The culturally-specific performances found at the Taj Mahal (Edensor 2000) are fascinating and well documented, and they are replicated in various forms at every cultural heritage site around the world. The heritage landscape provides space for heritage practices and performances, and *vice versa*. At Angkor Wat in Cambodia, every sunrise is greeted by hundreds of camera-wielding visitors who are simultaneously sight-seeing and performing the role of tourists, complete with the use of appropriate props. A visit to the Petaling Street market in Kuala Lumpur is infinitely less meaningful to tourists who do not play their part. The tourist role mainly involves adopting an attitude of faux-sincerity and haggling for goods that they didn't really want in the first place.

Cultural heritage also lends itself well to performances by workers. The Big Pit, a coal mining museum in Wales, UK, for example, features an underground tour with an ex-miner-turned heritage-actor. Bowen (2001) notes how the performances of chefs in floating restaurants in Malaysia are perhaps more salient to consumers than the quality of the food. One such floating restaurant in Kota Kinabalu has developed into a cultural heritage site in its own right. Kampung Nelayan (Fisherman Village) is a floating seafood restaurant, anchored on a freshwater lake. It features cultural shows depicting the music, dance, traditional dress, and simulated hunting of the Murut people. Few of the visitors watching and, at times, participating in the performances, are deterred by the fact that the Murut people are actually indigenous to inaccessible inland regions of Southwestern Sabah.

Whilst the performativity metaphor is extremely salient to the experience of cultural heritage, it is not without its problems. As Nash (2000) points out, although non-representational studies claim to be radically contextual, the non-representational is often elevated not only above the visual, but often the social and cultural world. The performances at cultural heritage sites are still 'mediated by words...scripted, performed and watched' (Nash 2000, 658), rather like a play. Visual and textual forms of representation constitute practices in themselves, and often occur simultaneously with acts of performance, yet there is rarely adequate room for them in non-representational studies. A further problem of relying solely on non-representational studies of cultural heritage concerns the ontological issue of analysing the pre-cognitive and non-verbal. There is a certain irony in producing texts of the non-representational, and a risk of conceptualizing individualistic 'sovereign subjects' rather than specific contexts. As Nash (2000, 662) argues, '...the energy spent in finding ways to express the inexpressible ... seems to imply a new (or maybe old) division of labour between academics who think (particularly about not thinking and the non-cognitive) and those "ordinary people" out there who just act.'

There is a danger, therefore, that non-representational studies are currently limited in their ability to address the dearth of experiential studies of cultural heritage. Embodiment, practice, and performativity are important within the experience of consuming cultural heritage, but they are only elements of the experience, often occurring simultaneously with visual practices. The visual and representational continues to play a significant role in mediating such practices and performances, and perhaps more importantly, we need to understand the subjective and intersubjective experiences of place consumers according to their own terms of reference. It would seem that neither representational nor non-representational studies alone are equipped to capture the 'existential dialectic' of cultural heritage experience (see also Selby et al. 2008). This implies beginning analyses from the perspectives of heritage consumers, and subsequently recognizing visual and performative characteristics of cultural heritage, should they be salient to visitors' experiences.

Perceiving

In *The Phenomenology of Perception*, Merleau-Ponty (1962) demonstrates how our experience of the environment is embedded in the way we settle ourselves in the world and the position our bodies assume in it...In many epistemologies, including the behaviouralism that influences many studies of cultural heritage, there is a basic dichotomy between the psychological and the physical world. The physical, objective world is conceptualized as given: data passes into the sense organs, where it is deciphered and the picture reassembled. Existentialists, however, have argued that it is possible to break away from this dualism of being and appearance (Sartre 1969), and focus on the monism of phenomena. Perceptions are conceptualized as existing in fields, and the contents of these fields are crucially influenced by how individuals focus their gaze and position their bodies in space. This depends on their interests at hand, as much as on the object itself (Spurling 1977). Each individual's unique perspective causes the experience of cultural heritage to be embodied, as 'images, sights, activities are all linked through the embodied motion of the observer...' (Crang 1997, 365). At cultural heritage sites, the interests and knowledge of the individual, combined with chance encounters and varying perspectives, ensures that no two individuals will have exactly the same experience.

According to phenomenologists, individuals are neither completely knowledgeable of their surroundings, nor psychologically constituting it. It is necessary to look at the 'whole mode of existence of the subject(s) in question' (Spurling 1977, 15). Of salience to the experience of cultural heritage, place consumers create an intentional arc that 'projects round us our past, our future, our human setting, or physical, ideological, and moral situation' (Merleau-Ponty 1962, xviii). Furthermore, it is this intentional arc that underpins intentional behaviour.

This is consistent with Heidegger's (1967) concept of 'Dasein' or 'Being-in-the-world'. Both Husserl (1969) and Merleau-Ponty (1962) differentiate 'operative intentiality' from the more conscious and deliberate 'intentiality of acts'. It is argued that the former is an important and neglected aspect 'of the world and of our life, being apparent in our desires, our evaluations and in the landscape we see more clearly than in objective knowledge (Merleau-Ponty 1962, xviii).

In the context of cultural heritage, therefore, we should be concerned with interests, motivations and identities of place consumers, as this has a significant influence on their experience. Perception 'arranges around the subject a world which speaks to him of himself (sic)' (Merleau-Ponty 1962, 132). It is common for tourists to cast a cultural heritage environment around themselves, by linking together various sites, landscapes, signs, symbols, and practices. The same locality may consist of several existential cultural heritage environments. The city of Liverpool, UK, for example, can be used by visitors to stage several different cultural heritage experiences. There is a maritime history environment, constructed by linking together elements such as the Albert Dock, the Merseyside Maritime Museum, the Mersey Ferries and the International Slavery Museum. Alternatively, visitors construct a Beatles landscape out of Mathew Street, The Beatles Story attraction and 'The Magical Mystery Tour'. As Kelly (1955) argues, the reality of an environment cannot reveal itself to us, instead it is subject to as many ways of construing it as we can invent. Kelly argues that individuals develop a system of constructions in order to understand and anticipate experience. The constructs used in anticipating events are believed to be both integrating and differentiating, so, by implication, cultural heritage is 'grasped not only as that which is, but rather also that which is not' (Schutz and Luckmann 1974, 173).

Knowing

Phenomenology avoids other dichotomies that hinder our understanding of cultural heritage experience. As Schutz and Luckmann (1974, 3) argue, 'everyday life is…fundamental and paramount reality'. In contrast to attempts at dichotomizing tourism and everyday life, as Bhabha (1994) suggests, seemingly mundane events and everyday actions are crucial in forming the identity of both people and places. Individual place consumers take some experiences for granted as common sense, whilst also sharing some experiences with others in an intersubjective world. The process of consuming cultural heritage draws upon a great variety of both immediate and mediated experiences, termed the 'stock of knowledge' (Schutz 1972). The stock of knowledge of an individual develops with new experiences, formed through both immediate and mediated knowledge. It is therefore influenced by *both* representations and practices (Selby 2004b). At cultural heritage sites, the visitor draws upon their stock of knowledge in order to give meaning to the exhibits, visual representations, and interpretive material. Without some intersubjective understanding provided by the stocks of knowledge of visitors, the experience would be devoid of meaning and emotion.

In familiar situations, individuals rely upon previously proven 'recipes' for acting, and these are likely to be socially transmitted. The performances noted at the Taj Mahal by Edensor (2000) are therefore underpinned by the recipes for acting provided within various cultural and social groups. This knowledge has been transmitted over time by parents, teachers, religious leaders, literature, the mass media, promotional materials, guidebooks, etc. Cultural meaning exists in linguistic typifications, yet 'in the natural attitude' people in the same social group take for granted these meanings. Individuals, therefore, belong to social systems divided by familiaral relationships, age groups, ethnicity, gender and social class, all of which influence the distribution of intersubjective knowledge. Place consumers partly experience cultural heritage individually, and partly though the positions and perspectives imposed upon them in an intersubjective 'natural world view'. No two individuals will have exactly the same stock of knowledge. However, cultural heritage tends to encourage both the development and use of, intersubjective knowledge. From childhood history lessons to tour groups at heritage sites, place consumers are significantly influenced by others.

Of significance to the debate regarding the visual nature of cultural heritage is the unproblematic way in which an individual glides between immediate and mediated experience. As Crang (1997) argues, the representations and experiences of tourism are closely interrelated, and should not be artificially separated. In the case of cultural heritage, the majority of the stock of knowledge will have been acquired through mediated sources (representations), and only when actually visiting a heritage site will individuals share the same temporal and spatial zones. From early childhood, both first-hand and mediated experiences are 'embedded in intersubjectively relevant, socially-determined and pre-delineated contexts' (Schutz and Luckmann 1974, 246–7). The style of lived experience determines what is taken for granted and what is problematic or new to the stock of knowledge. At a heritage site, it is when representations contradict the stock of knowledge that the visitor makes a conscious effort to confirm or reject the message through accumulating further information. This is similar to the process observed by Bagnall (2003) at Wigan Pier, UK. Bagnall also demonstrated how the memories and past experiences of visitors were drawn upon whilst consuming the site, through a process of reminiscence.

Authority figures (or opinion-formers) influential in the formation of mediated knowledge are themselves influenced by society's norms and values. Social groups and cultures also determine which knowledge is considered credible, relevant and communicable. Whilst the individual is free to form their own typifications in order to understand experiences, society removes the burden of so doing though the provision of language that reflects the social group and society's priorities. This process is more commonly understood and discussed in the context of the power conveyed by dominant discourses. Language therefore acts as a framework, even for first-hand experiences. In the case of mediated knowledge, or representations, the interpretation of knowledge is obviously detached from the original spatial, social, and temporal contexts. The interpretation therefore depends upon a degree of familiarity and empathy between the interpreter and the creator of the signs. It is important that 'the interpretation of signs in terms of what they signify is based on previous experience, and is itself a function of the scheme' (Schutz and Luckmann 1974, 119). The 'interpretive function' of signs, influenced by past experiences, can be very different from the 'expressive function' intended by the author. In the context of cultural heritage, it is important to evaluate both. The existence of social knowledge will ensure that there are significant variations in the interpretation of cultural heritage and its representations, and heritage managers need to be aware of these.

To phenomenologists, representations cannot be either true or false; just more or less plausible to a particular individual. This depends on the state of the individual at the time, and whether it contradicts or complies with other knowledge. It would seem that 'non-everyday knowledge' such as stories, myths, and the phantasmagoric can be influential within the stock of knowledge, as there are few competing representations to reduce the plausibility. Cultural heritage sites also exemplify what Schutz and Luckmann (1974, 264) term 'objectivation' – the 'embodiment of subjective processes in the objects of the everyday life-world'. Built heritage, in particular, provides the possibility of interpreting 'results of acts' (Schutz and Luckmann 1974, 271) that 'leave behind traces in lifeworldly objects' (Schutz and Luckmann 1974, 272). Again, this questions the distinction between the representational and non-representational. Despite this, the knowledge of cultural heritage is gathered in different 'provinces of reality', whether they relate to first hand experience, representations, or both. It is common for heritage landscapes to be delineated from the surrounding city-scape, in order to shift the place consumer into another province of reality. This will then influence practices that take place within that space. Cultural heritage sites, in general, create an unusual province of reality in that they combine first-hand experience with the active consumption of numerous representations. Many heritage sites create different provinces of reality through multi-media technology. The 'restorable' reach of the past, and the 'attainable reach' of the present can be facilitated through technology, and there has been a 'qualitative leap in the range of experience and an enlargement of the zone of operation' (Schutz and Luckmann 1974, 44). An implication of this is the apparent blurring between immediate and mediated experience. This is certainly the case in the National Museum of Singapore, where first-hand experience incorporates numerous multi-media installations and simulations, integrating and blurring physical and virtual worlds.

Acting

Instead of focussing on behaviour – a reaction to external stimuli – phenomenologists are concerned with action. Action is reflexive and intentional, even if the stock of knowledge may provide recipes for acting. This tends to happen in familiar situations where the likely consequences of the act are known. Recipes commonly followed at cultural heritage sites include acts such as taking photographs and following guided tours. The tourism industry attempts to add to the stock of knowledge by providing ready-made recipes. Promotional material produced by heritage sites often uses clichés (see Voase 2000), encouraging the visitor to 'step back in time' or 'trace the steps'. The use of cliché, of course, encourages the actor to draw upon their imagination (or knowledge), whilst avoiding the use of facts that might be contested. The provision of ready-made recipes for cultural heritage action is extremely profitable, as the popularity of *Lonely Planet* guides demonstrates. The stock of knowledge as a whole, however, determines 'the degree of freedom in the choice of various courses of life' (Schutz and Luckmann 1974, 95). The system of constructs that a person establishes represents the network of paths along which they are free to move (Bannister and Mair 1968, 27).

The stock of knowledge is also crucial to more deliberate acts of decision-making. In the context of cultural heritage, this includes the choices made by individuals concerning the information sought about cultural heritage, the decision to visit a site, decisions made at the site, etc. Through their stock of knowledge, place consumers are simultaneously involved in the consumption and production of heritage. As Anderson and Gale (1992, 4) point out, '...in the course of generating new meanings, and decoding existing ones, people construct places, spaces, landscapes, regions and environments...they fashion certain types of landscape, townscape, streetscape.' Franklin and Crang (2001, 16) perceptively point out that travel is finely integrated into everyday life, and is, for many, an activity that begins on cold dark winter evenings, when we produce 'a phantom landscape which guides our understanding of the one we eventually see'. A crucial part of this process is a form of 'map consulting' (Schutz 1970, 129). This involves both the 'reproduction' of memories and experiences, and 'retention', which is holding the picture before our inner eye. This process of anticipation, or projection, is based on the assumption that one can bring about a similar result as in the past though specific courses of action. Again, the stock of knowledge is believed to underpin these (visual) projections, as individuals place themselves at a future time when the act has been accomplished. The act is projected in the future perfect tense, and the actor becomes 'self-consciously aware of his phantasying' (Schutz and Luckmann 1974, 69). Deliberation occurs when there are overlapping or conflicting interests, and the individual isn't sure which knowledge is relevant. Each act then 'takes its turn in projecting itself upon the screen of the imagination' (Schutz and Luckman 1974, 69). Cultural heritage, because it usually involves travel and a visit that must be planned in advance, necessitates this rehearsal of action in the actor's imagination, rather more so than many other courses of action. Again, it would appear that this process is simultaneously visual and performative.

Concluding

The chapter began by drawing upon cultural geography and tourism studies in order to understand both the visuality and performativity of cultural heritage. It became clear, however, that although salient to the process of experiencing cultural heritage, any study deemed to be either 'representational' or 'non-representational' is likely to present a rather partial explanation. It is surprising that so few researchers interested in the experience of cultural heritage draw upon a branch of philosophy devoted to understanding how humans experience the world around them. Phenomenology offers valuable contributions in terms of understanding ways in which humans simultaneously view and perform cultural heritage, the importance of knowledge in experiencing cultural heritage, and the actions of cultural heritage consumers.

It will not have escaped the reader's attention that the chapter is structured through headings using the present participle or 'gerund'. This is intended to emphasize the active nature of cultural heritage consumption. Individuals help to constitute the cultural heritage environment that they consume by casting an 'intentional arc' around themselves. Cultural heritage consumers develop a stock of knowledge over many years, and they subsequently draw upon this stock of knowledge during the experience of cultural heritage. The stock of knowledge provides typifications that may be used to contest, as well as validate, exhibits and landscapes. Much of the stock of knowledge is acquired socially, leading to intersubjective experiences and practices. The stock of knowledge is also composed of both first hand experiences, and mediated knowledge (representations). Mediated knowledge is likely to form a much larger proportion of the stock of knowledge; and in the context of cultural heritage, most experiences will be mediated to some degree. Social groups and cultures provide recipes for acting that place consumers follow within culture heritage sites and landscapes. Finally, when the experience of cultural heritage does involve more conscious acts and decision-making, the likely consequences are projected visually in the future perfect tense, again drawing upon the stock of knowledge.

The introduction to the chapter contained a reference to the dialectics of cultural heritage experience – the symbiotic yet contradictory relationships between elements of the cultural heritage experience. Some time ago, phenomenologists argued that instead of dichotomizing the physical and the psychological, as many epistemologies insist on doing, experience is better captured through the 'monism of phenomena' (Sartre 1969). Phenomenology also offers hope in overcoming the numerous dichotomies blighting our understanding of cultural heritage experience. This implies the adoption of a much more holistic conceptualization of the experience of cultural heritage, emphasizing the 'monism of cultural heritage experience': people-place-past.

References

Anderson, K. and Gale, F. (eds) (1992), *Inventing Places: Studies in Cultural Geography*. (Melbourne: Longman Cheshire).

Bagnall, G. (2003), 'Performances and Performativity at Heritage Sites', *Museum and Society* 1:2, 87–103.

Bannister, D. and Mair, J.M.M. (1968), *The Evaluation of Personal Constructs*. (London: Academic Press).

Barthes, R. (1984), 'The Eiffel Tower' in *The Eiffel Tower and Other Mythologies*, trans. R. Howard (New York: Hill and Wang).

Beeho, A. and Prentice, R. (1995), 'Evaluating the Experiences and Benefits Gained by Tourists Visiting a Socio-industrial Heritage Museum: An Application of ASEB Grid Analysis to Blists Hill Open-Air Museum, The Ironbridge Gorge Museum, United Kingdom', *Museum Management and Curatorship* 14:3, 229–41.

Benjamin, W. (1939), 'Paris – The Capital of the Nineteenth Century', in W. Benjamin, *The Acades Project*, 14–26. (Cambridge, MA: Harvard University Press).

Benjamin, W. (2003), 'The Work of Art in the Age of its Technological Reproducibility', in M.W. Jennings (ed.) *Selected Writings*, 4, 251–83. (Cambridge, MA: Belnap).

Berjaya Hills (2008), 'Colmar Tropicale' [Online]. Available at: http://www.berjayahills.com/how.php [accessed: 17 November 2008].

Bhabha, H. (1994), *The Location of Culture*. (London: Routledge).

Bissell, D. (2009), 'Visualizing Everyday Geographies: Practices of Vision Through Time-travel', *Transactions of Institute of British Geographers* 34:1, 42–60.

Boorstin, D. (1964), *The Image: A Guide to Pseudo-Events in American Society*. (New York: Atheneum).

Bowen, D. (2001), 'Research on Tourist Satisfaction and Dissatisfaction: Overcoming the Limitations of a Positivism and Quantitative Approach', *Journal of Vacation Marketing* 7:1, 31–40.

Chhabra, D. (2008), 'Positioning Museums on an Authenticity Continuum', *Annals of Tourism Research* 35:2, 427– 47.

Cosgrove, D. (2008), *Geography and Vision: Seeing, Imagining, and Representing the World*. (New York: IB Tauris).

Crang, M. (1997), 'Picturing Practices: Research through the Tourist Gaze', *Progress in Human Geography* 21:3, 359–73.

Crouch, D. (2000a), 'Tourism Representations and Non-representative Geographies: Making Relationships between Tourism and Heritage Active', in M. Robinson, N. Evans, P. Long, R. Sharpley, J. Swarbrook, *Tourism and Heritage Relationships: Global, National and Local Perspectives*, 93–104. (Sunderland: Business Education Publishers).

Crouch, D. (2000b), 'Places Around Us: Embodied Lay Geographies in Leisure and Tourism', *Leisure Studies* 19, 63–76.

Culler, J. (1981), 'Semiotics of Tourism', *American Journal of Semiotics* 1, 127–40.

de Rojas, C. and Camarero, C. (2008), "Visitors' Experience, Mood and Satisfaction in a Heritage Context: Evidence from an Interpretation Center', *Tourism Management* 29:3, 525–37.

Degen, M., DeSilvey, C. and Rose, G. (2008), 'Experiencing Visualities in Designed Urban Environments: Learning from Milton Keynes', *Environment and Planning A*, 40, 1901–20.

Doel, M. and Clarke, D. (2007), 'Afterimages', *Environment and Planning D: Society and Space* 25, 890–910.

Echtner, C.M. (1999), 'The Semiotic Paradigm: Implications of Tourism Research', *Tourism Management* 20:1, 47–57.

Edensor, T. (2000), 'Staging Tourism: Tourists as Performers', *Annals of Tourism Research* 27:2, 322–44.

Foucault, M. (1977), *Discipline and Punish*. (London: Tavistock).

Foucault, M. (1986), 'Of Other Spaces', *Diacritics* 16, (Spring), 22–7.

Franklin, A. and Crang, M. (2001), 'The Trouble with Tourism and Travel Theory?', *Tourist Studies* 1:1, 5–22.

Gottdienier, M. (1995), *Postmodern Semiotics: Material Culture and the Forms of Postmodern Life*. (Oxford: Blackwell).

Halewood, C. and Hannam, K. (2001), 'Viking Heritage Tourism: Authenticity and Commodification', *Annals of Tourism Research* 28:3, 565–80.

Hall, S. (ed.) (1997), *Representation: Cultural Representation and Signifying Practices*. (Milton Keynes: Open University Press).

Heidegger, M. (1967), *What Is a Thing?* trans. W.B. Barton, Jr. and V. Deutsch. (Chicago: Henri Regnery).

Hewison, R. (1987), *The Heritage Industry: Britain in a Climate of Decline*. (London: Methuen).

Husserl, E. (1969), *Cartesian Meditations: An Introduction to Phenomenology*. trans. Dorion Cairns. (The Hague: Nijhoff).

Jamal, T. and Hollinshead, K. (2001), 'Tourism and the Forbidden Zone: The Underserved Power of Qualitative Inquiry', *Tourism Management* 22, 63–82.

Jokinen, E. and Veijola, S. (1994), 'The Body in Tourism', *Theory, Culture and Society* 11, 125–51.

Kelly, G.A. (1955), *The Psychology of Personal Constructs*. (New York: Norton).

McIntosh, A. and Prentice, R. (1999), 'Affirming Authenticity. Consuming Cultural Heritage', *Annals of Tourism Research* 26:3, 589–612.

Merleau-Ponty, M. (1962), *Phenomenology of Perception*. trans. C. Smith. (London: Routledge and Keegan Paul).

Morgan, N.J. and Pritchard, A. (1998), *Tourism Promotion and Power: Creating Images, Creating Identities*. (Chichester: John Wiley and Sons).

Nash, C. (2000), 'Performativity in Practice: Some Recent Work in Cultural Geography', *Progress in Human Geography* 24:4, 653–64.

O'Rourke, D. (1988), *Cannibal Tours*. (Cairns: O'Rourke and Associates).

Philo, C. (1992), 'Foucault's Geography', *Environment and Planning D: Society and Space* 10:2, 137–61.

Pile, S. (2005), *Real Cities*. (London: Sage Publications).

Poria, Y., Arie, R. and Avital, B. (2006), 'Heritage Site Management: Motivations and Expectations', *Annals of Tourism Research* 33:1, 162–78.

Prentice, R.C., Witt, S.F. and Hamer, C. (1998), 'Tourism as Experience: The Case of Heritage Parks', *Annals of Tourism Research* 25:1, 1–24.

Ringer, G. (ed.) (1998), *Destinations: Cultural Landscapes of Tourism*. (Routledge: London).

Ritzer, G. and Liska, A. (1997), '"McDisneyization" and "Post-Tourism": Complementary Perspectives on Contemporary Tourism', in C. Rojek and J. Urry, *Touring Cultures: Transformations of Travel Theory*, 96–112 (London: Routledge).

Rojek, C. (1997), 'Indexing, Dragging and the Social Construction of Tourist Sights', in C. Rojek and J. Urry, *Touring Cultures: Transformations of Travel Theory*, 52–74. (London: Routledge).

Sartre, J-P. (1969), *Being and Nothingness*, trans. H. Barnes. (London: Methuen).

Sauer, C.O. (1925), 'The Morphology of Landscape', *University of California Publications in Geography* 2, 19–54.

Schutz, A. (1970), *Reflections on the Problem of Relevance*. (New Haven, CT: Yale University Press).

Schutz, A. (1972), *The Phenomenology of the Social World*. trans. G. Walsh, F. Lehnert, (London: Heinmann).

Schutz, A. and Luckmann, T. (1974), *Structures of the Life-World*. trans. R.M. Zaner and H.T. Engelhardt Jr. (London: Heinemann).

Selby, M. (2004a), *Understanding Urban Tourism: Image, Culture and Experience*. (London: IB Tauris).

Selby, M. (2004b), 'Consuming Cities: Conceptualizing and Researching Urban Tourist Knowledge', *Tourism Geographies* 6:2, 186–207.

Selby, M., Hayllar, B. and Griffin, T. (2008), 'The Tourist Experience of Precincts', in B. Hayllar, T. Griffin and D. Edwards (eds), *City Spaces – Tourist Places: Urban Tourism Precincts*, 183–202. (London: Butterworth Heinemann).

Selwyn, T. (1996), *The Tourist Image: Myths and Myth Making in Tourism*. (Chichester: John Wiley and Sons).

Shrurmer-Smith, P. and Hannam, K. (1994), *Worlds of Desire, Realms of Power: A Cultural Geography*. (London: Arnold).

Spurling, L. (1977), *Phenomenology and the Social World*. (London: Routledge and Keegan Paul).

Thrift, N.J. and Dewsbury, J.D. (2000), 'Dead Geographies – And How to Make them Live', *Environment and Planning D: Society and Space* 18, 411–32.

Urry, J. (1990), *The Tourist Gaze: Leisure and Travel in Contemporary Societies*. (London: Sage Publications).

Urry, J. (2002), *The Tourist Gaze: The New Edition*. (London: Sage Publications).

Voase, R. (2000), 'Explaining the Blandness of Popular Travel Journalism: Narrative, Cliché, and the Structure of Meaning', in M. Robinson, N. Evans, P. Long, R. Sharpley and J. Swarbrook, *Expressions of Culture, Identity and Meaning in Tourism*, 413–24 (Sunderland: Business Education Publishers).

Wylie, J. (2006), 'Depths and Folds: On Landscape and the Gazing Subject', *Environment and Planning D: Society and Space* 24, 519–35.

York Archaeological Trust (2006), 'Group Visits Leaflet' [Online]. Available at: http://www.jorvik-viking-centre.co.uk/ [accessed: 17 November 2008].

Chapter 4
The Perpetual Performance and Emergence of Heritage

David Crouch

This chapter develops a discourse on heritage that seeks to achieve two things. First, it seeks to dislodge 'heritage' from the conventional concept of its being somehow pre-figured or ready-made. Second, it positions heritage beyond the visual and, whilst holding on to the visual as a component in the dynamic constitution of heritage, acknowledges it to be one among many commingling energies. These energies are discussed through a consideration of performance and performativity, the spaces of heritage, and the flows of influence inherent in its continual making and emergence. The relationality, rather than polarity of representations and what is not representational, is examined through recent debates on performativity. Heritage is related to cultural identity, feelings of belonging and the play of memory and duration. The character and meaning of heritage may be influenced by representations, including those that are visual; heritage is continually emergent in living.

I have long been uneasy with the notion of 'heritage'. It tends to be used to fix particular assertions about what heritage is, materially or metaphorically, and to provide a script for what it means. Moreover, it can essentialise in a way that is increasingly problematic. The problem is not with heritage, but the way it is thought about and institutionalized in contemporary culture, often through dominant visual representations. Heritage speaks of Great Things that become objectified and often reified: Great Places, Sites and People. The objectified becomes distanced from the real, the everyday, the experience of individual lives. Heritage is familiarly written in terms of heritage landscapes that prompt the constitution of those 'landscapes' that comply with its prefigured, implicitly judged elite and its priorities (Moore and Whelan 2007). Of course, there are other heritages, most notably of Indigenous peoples, their livelihoods and their land.

The character of elite thinking is shared across much of the world, particularly in Europe, in North America, and in many countries of the developing world. Significantly, there is a paradox at work here. Heritage is distanced and detached, whilst at the same time it is identified and found in communication as a component of belonging. Heritage becomes signified; produced and constituted in cultural contexts; communicated in cultural mediation; consumed, further reified, and 'held onto' as a sense of belonging. Heritage is, by such means, ritualized in cultural practise inscribing a particular world view that is circulated in mediated popular culture. In this chapter, I want to go beyond the more familiar heritage critique in terms of selectivity, elitism and politics, with which I profoundly agree, to the more nuanced, personally varied issues regarding the dilemma of thinking heritage-cultural identity, and what I outline as gentle politics that emerges from the quieter affects of people coming to their own heritage.

In the 'western world', heritage is habitually communicated and celebrated in the media, for example in tourism, in mediations of cultures of cities and countryside areas, even of 'nature'. Across these mediations, heritage comes across as prefigured; largely fixed. This presumption of being prefigured is part of a wider notion of culture as prefigured, contextualized, for example, in cultural capital. Yet the field of cultural studies has progressed, in ways related to cultural geography and social/cultural anthropology, as more processual, complex, changing, in flows and sometimes unexpected (Crouch 2009a).

Space is not merely understood and held through one dimensional reference to visual material. The visual is in any case not merely a practice of detached gazing, no matter how much subcategories have tried to permit variation within the frame. Moreover, the visual, as representations, tend to be thought of as life-detached objects. Heritage is thus embedded in an 'Other' sphere of cultural existence: heritage is pre-figured by others, pre-determined, as ready-made then made selectively available. What we mean by the visual includes a caring, caressingly haptic character and relationship with others, with the non-human (plants, a cat) and things, such as landscapes. The visual is constitutive, a process of creativity. In this chapter, it is considered as working relationally with other senses, sensations, thought and feeling. It is therefore necessary to refer the merely visual to a much broader process of life.

If we remove heritage from its pre-figured detached character it is possible to reconstitute the process of heritage. This can be done through an engagement with performance and performativity; concepts of culture and cultural production; and such related notions as belonging. But before engaging these productive components of process I turn to the question of time and heritage. Heritage tends to be considered in relation to a particular, 'sufficient' distance of the past (from the present). It is appreciated that contemporary events can remake the past, its memories and heritage. I want to go on to argue that heritage is not only constantly in the remaking – through, for example, festivals, the use of particular identified heritages in advertising, re-associated with new products and so on – but is always emergent in the present. Yet this re-emergence in no sense denies the potential relationality of 'new' heritage and other, pre-existent and also emergent contemporary pasts and their heritages.

Heritage is *represented* in numerous visual ways. Like visual culture more generally, it exists in buildings, forms in landscapes, clothing and fashion, ritual and popular culture more widely. It exists in family photograph collections and souvenirs, ornaments, letters and much more. Heritage is mediated in popular culture through many of these and in advertising; where something of heritage value is advertised, or may be used to advertise something else. Heritage associates cultural capital with institutional and corporate identity.

Place, or space, is familiarly a leitmotif of heritage and through which significance in heritage is conveyed and communicated in media: space as place, landscape, site and so on. That space, or heritage as space, is habitually conveyed in popular and specialist magazines, literature and film, and in tourism content through visual images (Rose 2001). There is a risk of over-emphasizing, indeed of privileging, the visual and its representations that I argue becomes explicit through a critical consideration of some years of conceptual debate and empirical evaluation. Too frequently, claims of the power of the visual have been grounded in an absence of empirical evidence, or, when such evidence has been available, it has been sourced merely in relation to visual culture. This problem is evident in pervasive emphasis on the visual with regards to tourism, particularly through the notion of the gaze, and as applied to heritage (Urry 2003). Curiously, Urry (2002) in particular has persisted in referring back to a Renaissance 'noblest of the senses' presumption.

Until recently, tourism advertising tended to emphasize the visual, and to a degree it still does, although there is an evident increasing use of non-visual methods (Kirshenblatt-Gimblett 1998). In the early 1990s, cultural studies and cultural geography, for example, had also focused their concerns upon representations that were often mainly of a visual character (Duncan and Ley 1991). In cultural studies, for example, the body emerged as object; an objectification of the gaze (Featherstone and Turner 1995). In geographies of landscape, the latter emerged as a tract of country ordered in a masculine vision (Rose 1993). In social anthropology, emphasis has typically been less restricted to visual representations because of its disciplinary interest in what people do, yet the signs of rituals, for example, have tended to be focused as opposed to what those participating felt and thought in the doing. As Taussig argues, the features and gestures of performance are operated not of 'sense so much as sensuousness, an embodied and somewhat automatic knowledge that functions like peripheral vision, not studied contemplation, a knowledge that is imageric and sensate rather than ideational' (1992, 141).

The last two decades have disrupted the habitual privileging of the visual and its representations, not in a way to reject their role but, rather, to adjust our reading of their significance in meaning and identity. Moreover, representations are not merely of a visual kind; sound, taste and all the senses have the capacity to represent. Furthermore, visual culture is also tactile: in buildings, in souvenirs and so on. Each of these varied components affecting and occurring in heritage do not act in isolation but interplay, act in mixture. Discussion on visual representations, exemplified in Urry's concentration on heritage in his tourist gaze, presumes a linear association of heritage in, and as, visual representations that frame experience. Often, thereby, visual representations of heritage, or a version of it, are presumed to frame, or prefigure the value accorded to a particular heritage. As a consequence, heritage becomes mediated and consumed in an unproblematic way. Yet important developments over the past two decades provoke a more critical thinking, and plunge that simplistic association into a complexity of how heritage is constituted, composed, and also how the dynamics of its communication work. To do this, it is necessary to dismantle the privileging of visual culture in understanding heritage and the 'experience' of heritage. An example of experiencing heritage in ways that include its representations is in the embodied wandering around, across, by or amongst a space that is, visually and otherwise, associated with heritage. This requires attention to recent innovative discussions concerning space. But first I examine the relationship between, or perhaps amongst, representations and what is not representational. Significantly this distinctive labelling dissolves.

Rethinking Heritage's Representations Dynamically

In a recent challenging discussion of artwork, Barbara Bolt (2004) examines the notion of 'art beyond representation'. She notes that Heidegger talked of the power to go beyond; that art is more than the intention (2004, 185). She finds in Deleuze a means to address and engage the character of what she calls 'working hot' in the performativity of art in ways that I will argue informs our grasp of the flows of heritage: 'In rhyming the rhythms of the landscape and the body, meaning and reality are constituted in performance' (2004, 171 et seq).

In a fascinating discussion of the 'material productivity of the performative act', Bolt addresses the performed materiality of plastic arts. Taking the idea of performativity in and of artwork further, she argues that representations continue to participate in flows of poetic possibilities in their public availability. The performative 'life' or vitality of the artwork – even two dimensional works – is performed, too, by the individual in his and her encounter with it. Two dimensional pictures may not be experienced only through the gaze, but with diverse dispositions of the body, memory recall, inter-subjectivity, emotion, fear and anxiety, unlike the formal viewing mistakenly associated with art in the gallery (Jones 1997). Despite the apparent limitations, or confinement as a completed and available object, in its object-ness the painting continues in flow. Bolt's thesis offers a critical intervention in conceptualizing the working of landscape that moves forward the critical conceptualization of art as beyond representation. Recent contributions on the merging of art and everyday life further prompt engagement and relationality between modalities and realms of performativity in terms of landscape (Cant and Morris 2006).

I want to urge that we reconsider 'heritage' as available in visual representations outside such limitations, and reconsider and refigure it as alive in them. David Matless (1992) noted that representations are the product of living and emerge through practice. Pictures, visual representations, are not isolated from the world we all live in. They have no 'priority' over making sense of things, what they mean, how things and experience relate, if at all, with any of our identities. They are merely 'available' in our lives. They have no priority or privilege over our lives and what we make of them, or things and events we engage and may encounter. Instead, the *liveliness* in representations, rather than their fixed status, however deadened in their commercial misapplication (for example, taken out of context), is open to commingling in the multiple flows of vitality that the individual finds her/himself amongst; a wider agenda. This is the wider agenda in which heritage waits to be understood.

In this chapter, I pursue these questions and concerns through attention to process rather than category (representation, life, etc), drawing primarily on recent interventions in art theory. Articulating what heritage *is*, rather than how it emerges and happens, feels very incomplete, as Tolia-Kelly (2008a,b) acknowledges in a recent discussion of varying interactions with English landscape and its cultural resonances. If it is that we live space, not merely in relation to it, there would seem to be more going on than an evocation of mental cultural resonance. At the core of this feeling of incompleteness is a sense that heritage, representations and space need, therefore, to be conceptualized relationally. The work done by representations is done through a complexity of life, performance, memory: the privileging of visual culture is thus hugely reductive. I argue for considering the construction and constitution of heritage through this complexity, where the work of representations *vis-à-vis* heritage might emerge much more informed and more complex. The power of the gaze becomes unsettled, disrupted, and visual representations – and how the visual works – must be considered 'in the mix'; relationally.

Representations and Spaces of Heritage

Recent critical attention to space directs attention to its relational, dynamic, contingent character. Space emerges from this as persistently 'in the making', through a complexity of forces, influences and practices. Massey's (2005) focus on space (itself) as relational to flows, energies and things, renders space closer to the lived and human. She articulates the character of space as relational through the connectedness and dynamics of things. Significant components of her thesis are that space is produced of inter-relationships in life, and that therefore space is always under construction, in flows of influence, in process. Casey (1993) has written of more explicit connections of space to life, regarding them as mutually constitutive.

British geography's recent interest in the French theorists Deleuze and Guattari (2004) has focused upon their notions of territory, space and spacing. In geographically pertinent terms, this space is highly contingent, emergent in the cracks of everyday life, affected by a maelstrom of energies well beyond human limits. What interests them is the potential of space to be constantly open to change and becoming, rather than only or mainly as the more settled. Interpreted in terms of individuals' participation in space, in making space through spacing, space and life cohabit in holding on to the familiar and going further into what is unknown. Adjustments are produced through which life can be negotiated, always in tension in an unlimited array, or immanence of possibility. Moreover, life does not work to a given script or prefigured world, not even linearly with our own memory and its spaces. To a degree this process happens in an embodied way; it could not do otherwise, as Deleuze insists upon the possibility of everything being involved in this process, not merely mental reflexivity, and potentially including memory of other times and spaces engaged relationally. Spacing happens in highly intense and in less urgent moments. As Deleuzian geographer Bonta (2005) argues, this awkward and pregnant vitality enables us to move beyond landscape as (fixed) text; or, we might infer, heritage as framed by representation. These arguments emerging from Deleuze and Guattari have come to be labelled 'non-representational' theories (Thrift 2008). Instead of working off representations of things, their debate emerges in the liveliness of things.

As Matless (1992) observed in the early nineties, visual and other representations are with us, not detached there from. Perhaps paradoxically, the so-called 'non-representational theory' is not antagonistic to representations, but regards them as fluid and engaged. I suggest in this chapter that understanding the performance of representations both in the making and in its mutual articulation in life beyond its making may offer a means both to deepen an understanding of heritage in relation to life, and space in relation to living. Thus, as Dewesbury et al. (2002, 473–9) explain with regard to representations and the thinking influenced by Deleuze and Guattari in particular:

> [W]e want to work on presenting the world, not on representing it, or explaining it. Our understanding of non-representational theory is that it is characterized by a firm belief in the actuality of representations. It does not approach representations as masks, gazes, reflections, veils, dreams, ideologies, as anything that is laid over the ontic [for me, life and its meanings]. Non-representational theory takes representations seriously … not as a code to be broken or as illusion … rather apprehended as performative in themselves; as doings.

From this discussion emerges another way of thinking and intellectually apprehending the way that visual representations of heritage 'sites' construct landscapes of heritage. In experiencing a heritage 'site' we engage in a process of spacing, with its openness to possibility, disruption, complexity, vibrancy and liveliness. Heritage is constituted in being alive. Heritage is situated in the expression and poetics of spacing: apprehended as constituted in a flirtatious mode: contingent, sensual, anxious and awkward. Of course there are visual representations of heritage that are of numerous and diverse kinds. These commingle in the greater complexity of flows and heritage is thus performed.

On Consumption, Representations and the Visual

The character of visual representation and heritage as consumed offers another means to unravel the complexities at work, in particular the work of social anthropologists such as Daniel Miller, and cultural theorists as far back as Raymond Williams and John Fiske (1989, 1990). Miller's (1987, 1997, 1998) work excavates the ways in which individuals refigure and thereby re-value things they buy in ways that change the prefigured and familiarly visual represented character, value and meaning with which producers enframe them. Miller has studied DIY kitchens in north London through to the drinking of Coca-Cola in Trinidad, unravelling the dynamic processes through which things, material objects, matter, identifying the production of meaning through consumers' practice. The too familiar polarization of popular culture as production and consumption has long been disrupted in the work of Williams (1989), who argued that culture was constituted in ordinary, everyday life; in relation to who we are and what we use and do. Fiske (1990) brilliantly drew out the practice of this process in his examination of the way a pair of jeans is given meaning and value in life. More recently, Sara Cohen (2007) disrupts the notion of what makes a City of Culture as she acknowledges the continual cultural and creative activity amongst small groups of people across the city in their own activities, making music and knitting. In a similar way, making a risky jump, heritage may matter. Interestingly, in a recent paper Gudrun Helgadottir (2008) has elucidated the role of knitting in the contemporary signification of Iceland's heritage. Knitting is alive in contemporary life as well as in the heritage 'image' of Iceland. The contemporary revival of knitting, she argues, offers an example of how heritage as projected and mediated can flow amongst heritage both ongoing and experiencing renewal.

In this work, there emerges a sense of what psychology theorist Rom Harré calls 'the feeling of doing' (1993). This unsettling of the visual across many disciplines disrupts the notion of the power of the visual sign. The power, or effect, of the visual can only be considered in relation to, relationally with, this greater way in which life happens. Burkitt (1999), from a perspective of embodied practice, argues its importance in the refiguring of negotiations of the complexities and fluidities of contemporary identities and values, and argues that 'the bodily' is significant in the ways that identities are constituted. The re-constitutive suggests a potential refiguring of prior, and given, identities and identifications in the contextuality of representations. Identities may be constituted and characterized in practice and performativity, and negotiated with contexts. Through our bodies, we expressively perform who we are. Heritage, however it is experienced, engages in a wider self, and a wider self-other, relationally. 'Doing heritage' may thus engage, amongst other things, visual representations, but it is also most likely to be considered and reflected upon in relation to other things that matter in life.

In his book on national identity and popular culture, Tim Edensor (2002) elucidates how identity is refigured, and made, through activities, events and what people do. Rather as Taussig argues, as we need to go beyond presumptions of prefigured meaning in the practice of rituals, so Edensor attends to life through which heritage may be grasped, understood and explained. He points to popular events on a national and very local scale. It may be that, if engaged through living, events, dwelling and simply doing things, whatever heritage is may become significant. What may not be engaged thus may not become significant, at least not in a popular manner.

Heritage, Dwelling and Belonging

A more vital conceptualization of heritage is possible through acknowledging heritage as closely engaged in dwelling, identity and belonging. Moreover, heritage is familiarly associated with 'time': things older than our own life tend to be categorized as heritage. Heritage can also be emergent in, and constituted in, our own lives and related to cultural identity, and it thus becomes possible to consider the work we do in possibly connecting such heritage with other institutionally categorized and mediated heritage. So far, we have considered how this may happen in relation to particular heritage-labelled 'sites'. Such heritage is relational amongst cultural identities, families, and places. It is valuable to consider this vitalism of heritage through a particular attention to the dynamic processes through which heritage emerges at particular times, moments, durations and feelings of belonging. In pursuing these ideas, I draw attention to a particularly interesting way in which to understand the negotiation of life and where we may find heritage being signified in understanding a process of holding on and going further. Rather than serially and linearly progressing our lives, our lives move in flows. In deciding what to do or where to go we negotiate [in] the tensions of life between holding on to an identity and 'going further', making adventure.

Tim Winter's (2007) fascinating investigation of (tourists) visiting Angkor Wat in Cambodia engages the ways in which individuals moving around a world heritage site do so in an embodied way. Here, meanings, based on a dynamic *heritage-in-the-now*, with memory, identity and its politics, emerge through the diverse ways in which individuals, collectively and differentially, encounter these sites in doing, feeling and thinking in the process of visiting. Winter situates his ethnographic investigation of the meanings that individuals and families visiting Angkor Wat constitute from this site in relation to the hard political history that much of Cambodia, including the area concerned, experienced in the latter half of the twentieth century. He acknowledges that this distinctive history has made an ancient site like this particularly important, as a pilgrimage for Cambodians. Identity and belonging are important, core inflections in this pilgrimage. Yet even in such a severely identified site, the mode of engagement – and the character and colour of the experience of visiting – are not easily predicted. Indeed, one of Winter's informants expresses this very multiplicity:

> [W]e come every year to see Angkor Wat, to have fun with the family, to see lots of people. We just drive around … We don't know about the history so we like to picnic here [at Angkor Wat] and [the reservoir]. These places we have met most people, the other places are too quiet. We like to see it crowded, both with foreign people and Cambodian people. No matter how busy we are, we come here over new year, we feel we have to come to see the people at this time (Winter 2007, 141).

As Winter asserts, 'clearly near at Angkor is an example of a tourism practice, whereby "the activity and its spaces are enlarged in the imagination"' (Winter

2007, 141, quoting Crouch 1999, 271). Rather than a constrained, visual or other-representational focus upon what their heritage is or must be, these individuals, even with the enormous power that Angkor Wat conveys, are contributing to the making of their own heritage. Whilst Winter records others who participate firstly in a more expected celebration of a particular site in a particular heritage, identity and belonging, this is by no means universal. Heritages are complex, fractured and variant. Moreover, heritage is unpredictable and does not work according to a given script, in visual representation or otherwise.

Although we 'may retain no trace of the temporal dynamic of the flow of time' (Bachelard 2000, 57), moments of performance when and through which things are remembered as significant can be revealed. In the doing, moments of memory are recalled, reactivated in what is done, and thus, while memory may be drawn upon to signify, it is less that memory is performed than that it is 'in performance'. Thus memory and the immediate are performed as complexities of time. The informants do not simply remember by picking the memory up momentarily, they return to it through performance and reform it. Time, too, is performed again and again, differently, and embodied thereby, grasped from clock or other time and instead wound up in body-performance.

Culture can be inscribed on the body surface in the wearing of adornments in tourism experiences: at a disco, surfing, or at a Balinese festival. The body is also a means to express oneself, how one feels, how we relate to each other and the world. Thereby the 'fun' of tourism may be a means of being in the world, of reaching and engaging the world, a medium through which it is enjoyed, and the subject declares herself within that world. Dance, white-water rafting and other 'adventure tourism' (Cloke and Perkins 1998), but also more mundane tourism such as camping and coach-touring and moments of emerging from the bus to stand and stare, provide exemplars that express not only systems of signs but expressions of feeling, subjectivity in the world, and our unique personality. Spatial practices provide means through which the individual can express emotional relationships with others and make their own sense of what is there. The ways in which the individual's body encounters space, events and activities informs the character of the encounter she or he makes, consuming place through her own embodied agency and making meaning in what is done and through the ways in which it is done.

Sociologist Ann Game (2001) argues, through her own reflections on being a visitor to Bondi Beach and to the English Pennines, that the individual constructs her own 'material semiotics' of places through the bodily encounters that she makes. She places the image of Bondi for surfing and its brochures to one side as she identifies the components of embodied practice that inform her meaning of this place. She may surf, but her meanings of the beach, that has an existence outside tourism 'products', are constituted of feeling the sand and her body pushing through water; memory of reading book whilst lying down, refigured in habitual reading moments on the beach.

In making this material semiotics come alive in the embodied human encounter with space, as in the case of heritage, I have suggested the idea of embodied semiotics (Crouch 2001). A reading of the visual signs needs to be reconfigured in terms of this practical ontology. Semiotics is not a visual but a physical and mental energy, both wholly commingled with the body's activity through which so much in terms of impulse, rather than rational information, for example, is felt. As soon as we acknowledge this more complex dynamic of grasping and making, to an extent, some sense of the world, the notion of linear and privileged influences of the material of vision and its projected semiotics is unsettled, to say the least.

Belonging often involves a nostalgic characterization of the past. In contrast Ann Game (2001, 227), more recently, argues for a belonging that is experienced in our everyday living yet not divorced from memory and which emerges out of feeling like a child:

> Moments when we feel wide-eyed, wide open, in love with the world. Running into the waves, the salt-smell spray in my face, or feeling the sand between my toes … these are moments of feeling 'this is right', 'now I have found what I have always been looking for, what I have always known'. I get that 'coming home' feeling … that might best be described as a sense of belonging.

Belonging emerges through the duration of life, and in its journeys, both momentary and over a longer trajectory; they produce flows of time that can be detonated into significance, in the way that Grosz later developed the potential of performativity (1999). Journeys in life happen in durations, across time and times. The philosopher Henri Bergson (1964) identified duration as the character of time in its continuous progression and heterogeneity. That duration is felt in new experiences and feelings impacting upon the past, rather than merely being added sequentially at the same register. Even habitual practices impact and change the character of what was done in the same habitual practice the previous time (Bergson 1964). Time thus emerges as relational, not preset or empty. Bachelard (2000) argued that 'instants' of time became 'fleshed out and filled in later'. Thereby, whilst acknowledging a thread of continuity he engaged a complexity of heterogeneity, shuffling, refiguring of time in memory: diverse, vibrant and full of energy. Time becomes revised and revisited and moulded as our journeys move. The general notion of mobility is more of an openness to possibility and the complication of time, a 'qualitative multiplicity' (Bergson, cited in Harris 2005, 42). Our own experience of things complicates prefigured, given notions of heritage.

Thus heritage in all its forms, great views and intimate corners, adventure and theme parks are brought into our lives, and may not remain detached from our own identities. To an extent, we encounter these in much more diverse, nuanced, and potentially intimate ways, re-engaging memories of our own and stories of others we may know, modulating our distinctive culture with the performance of the space; connecting, working mediated significance into our own story and into our own negotiation of life (Jewesbury 2003). Even in grand sites, the tourist performs in close encounters of performance too (Edensor 1998, 2001). This means that artefacts, heritage features and its landscapes are embodied, socially, through our relations with each other.

A consideration of 'heritage' in relation to English package tourists in Magaluf might appear to be absurd. However, in a deeply investigated and shared experience of time spent with a particular group of package tourists in Mallorca, Hazel Andrews (2004) identifies an appeal to identity. She documents a range of activities shared amongst individuals who by and large did not know each other before their visit. Those activities are, she argues, significantly racist, sexist, homophobic and nationalistic. A likely excess of alcohol and cross-over references between food and sex, and food and nationality (including the ubiquitous all-day English breakfast), combine with these other characteristics in constituting a sense of belonging, and an English identity that the individuals feel is difficult to access in England itself. Their imagined identity is with a distanced, but fantasy past of national heritage. Their heritage is both evident in visual representations of things they engaged, but also lived in what they did, in clubs, pubs and in how they disported themselves. The visual images resonate with their own behaviour; they remind them they are 'at home' [sic].

Heritage is frequently much more physically and visually amorphous, or nuanced and complex, than sites and representations, at least those that are mediated. One of the greatest reasons individuals visit somewhere away from home is to visit friends and relatives. We visit across town, across country and across continents. Visiting participates in flows of living, rather as Lynne Pearce (2004) captured her own complexity of feelings:

> ... travelling 'back home' is always, of necessity, a journey thru home as well as space. Things are changing- in the house, in the surrounding villages ... my parents are ageing. My returns blind me to a good deal of this. My apprehension of this home, indeed, has all the qualities of a dream where past and present mix and coalesce ... my parents are seen and remembered as they were; sometimes they are now ... My own ghost, meanwhile, flits around the place in a state of intermittent erasure.

These words translate very well in the idea and feeling of heritage: significances, meanings and feelings that are substantially of and through our own lives. Visiting friends and relatives contains, and can be signified in the sharing of stories, listening to sounds, touch; the celebration of passages in life: children, death. There is visuality – remembered features of expression in faces; photographs, touch and tension. These mix.

I worked on a photographic documentary of people's lives in an upland part of the north-east of England with a professional photographer, Richard Grassick of Amber Films Collective (Crouch and Grassick 1999; Crouch and Grassick 2005, 42–59). The project was provoked as a reaction to prevailing visual images of the locality's heritage at the time. These images projected to tourists and other visitors the landed estates and houses of the historic wealthy landowners. To us this seemed a limited lens onto heritage. Our particular interest was to explore a different heritage, located in individuals' lives and relations, where they live, feelings about the place and their life there. We documented five loose categories of people's lives and their stories: mothers working in a local factory and collecting their children from the leisure centre, people whose families have lived in small holdings of land for generations and new arrivals from other parts of England; people working at the quarries. We produced visual representations and text to combine place and feeling of experience. In the resulting work individuals were encouraged to be alert to the nuanced character of lives, smells and hidden stories; the story of how the ground is worked; to feel the air, as well as the land with both feet, as much as looking around themselves.

In 1991, I was asked by Channel 4 TV to write a book about community gardens or allotments as they are known in the UK. At the end of the short book, I included a 'heritage trail' by way of a listing of 12 allotment sites: 'Allotments are not places on the Heritage Trail, but might well be. Many of them have distinctive landscapes rich in their diverse use of materials, decorative in their crops and welcoming in their people' (Crouch 1991, 13). Allotment heritage is constituted in the working of these plots of ground, often inter-subjectively amongst others and as a site of belonging, to use Ann Game's words.

The reworking of heritage is exemplified in allotments in several ways. Their familiar popular heritage in the middle of the last century was of tired pieces of land and poor, inefficient, anachronistic people. A decade earlier, they had been a sparkle of national identity as they came to be reservoirs for food production during the Second World War (Crouch and Parker 2003). More recently, they re-emerge as sites of multiple identities, often multi-ethnic, but sometimes refuge for refugee groups where their distinctive heritages of cultivation and the position of cultivation in everyday life is sustained, 'cultivated' today. These sites also have new cultural identities, too, across cultural capital, in, for example, the ecological movement. One example of the latter in drawing upon an earlier version of its heritage is found in the reuse of the Second World War's allotment symbolism 'dig for victory', marked by a foot pushing a spade into the earth, widely distributed in posters and seen in cinemas at the time. During the 1990s, the poster was restyled as a banner against commercial development of these sites, using the term 'dig in for victory' marked by a bent eco-warrior keeping on digging.

Conclusions

Together with time and memory, heritage is always emergent, in process:

> I have become a collector of shards. Shards of memory, things passed down: told to me at the end of this long line of telling. I want to catch these shards, this half-lit, often, paste jewels. I don't know how authentic they are, does it even matter? For me it doesn't matter. I am making anew, building something from

> the remains. Wanting to honour the fleeting; the fragment, fractured histories and stories. Not passed down, but dredged up (Terri-Ann White 2004, 520).

White's grasp of heritage has echoes in the evidence, as well as the conceptual argument, that this chapter has followed. Heritage is open, full of possibilities, and never reliably constrained, however much the self-publicity of such as the tourism (rather than tourist's) industry asserts. Heritage is too rich to be so constrained. Heritage is not deferential; it is in and of life.

The process of constituting heritage that I have sketched in this chapter I will call *heritaging*. It resembles the complexity of what is called pilgrimage. Taking the example of Chaucer's *Canterbury Tales*, it is evident that the pilgrimage was a vehicle for much more than what might be realized as sacred, spiritual or religious experience. Drinking, flirting and much more happened on those journeys. The participants constituted their heritage along the way, in relation with the idea of being Pilgrims, and their desires that bore them to and back from the event in a longer journey of life and its heritages.

References

Andrews, H. (2004), 'Escape to Britain: The Case of Charter Tourists to Mallorca', unpublished PhD thesis, London Metropolitan University.

Bachelard, G. (2000), *The Dialectics of Duration* (Manchester: Clinamen Press).

Bergson, H. (1964), *Creative Evolution*, trans. A. Mitchell (London: Macmillan).

Bolt, B. (2004), *Art Beyond Representation: The Performative Power of the Image* (London: I.B. Tauris).

Bonta, M. (2005), 'Becoming-forest, Becoming Local: Transformations of a Protected Area in Honduras', *Geoforum* 36:1, 95–102.

Burkitt, I. (1999), *Bodies of Thought: Embodiment, Identity and Modernity* (London: Sage).

Cant, S.G. and Morris, N.M. (2006), 'Geographies of Art and the Environment', *Social and Cultural Geographies* 7:6, 857–62.

Casey, E.S. (1993), *Getting Back into Place: Towards a Renewed Understanding of the Place-World* (Bloomington: Indiana University Press).

Cloke, P. and Perkins, H.S. (1998), 'Cracking the Canyon with the Awesome Foursome: Representations of Adventure Tourism in New Zealand', *Environment and Planning D: Society and Space* 16:2, 185–218.

Cohen, S. (2007), *Decline, Renewal and the City in Popular Music Culture: Beyond the Beatles* (Aldershot: Ashgate).

Crouch, D. (1991), *The Allotment: A Viewer's Guide* (London: Channel 4 Publications).

Crouch, D. (2009), 'The Diverse Dynamics of Cultural Studies' in Jamal, T. and Robinson, M. (eds), *The SAGE Handbook of Tourism Studies* (London: Routledge).

Crouch, D. (2010), 'Flirting with Space: Thinking Landscape Relationally', *Cultural Geographies* 17:1, 5–18.

Crouch, D. and Grassick, R. (1999), *People of the Hills* (Newcastle: Amber Films).

Crouch, D. and Grassick, R. (2005), 'Amber Films, Documentary and Encounters' in Crouch, D., Jackson, R. and Thompson, F. (eds) *The Media and the Tourist Imagination: Convergent Cultures* (London: Routledge).

Crouch, D. and Parker, G. (2003), '"Digging-up" Utopia? Space, Practice and Land Use Heritage', *Geoforum* 34:3, 395–408.

Deleuze, G. and Guattari, F. (2004), *A Thousand Plateaus* (London: Continuum).

Dewsbury, J.D., Harrison, P., Rose, M. and Wylie, J. (2002), 'Enacting Geographies' *Geoforum* 33, 437–40.

Duncan, J. and Ley, D. (eds), (1991), *Place/Culture/Representation* (London: Routledge).

Edensor, T. (1998), *Tourists at the Taj: Performance and Meaning at a Symbolic Site* (London: Routledge).

Edensor, T. (2001), 'Performing Tourism, Staging Tourism: (Re)producing Tourist Space and Practice', *Tourist Studies* 1:1, 59–82.

Edensor, T. (2002), *National Identity, Popular Culture and Everyday Life* (London: Berg).

Featherstone, M. and Turner, B.S. (1995), 'Body and Society: An Introduction', *Body and Society* 1:1, 1–12.

Fiske, J. (1990), *Understanding Popular Culture* (London: Routledge).

Game, A. (2001), 'Belonging: Experience in Sacred Time and Space', in J. May and N. Thrift (eds) *Timespace: Geographies of Temporality*, 226–39 (London: Routledge).

Grosz, E. (1999), 'Thinking the New: Of Futures yet Unthought', in E. Grosz (ed.) *Becomings: Explorations in Time, Memory and Futures*, 15–28 (Ithaca, NY: Cornell University Press).

Harré, R. (1993), *The Discursive Mind* (Cambridge: Polity Books).

Harris, P.A. (2005), 'To See with the Mind and Think Through the Eye: Deleuze, Folding Architecture and Simon Rodia's Watts Tower', in I. Buchannan and G. Lambert (eds) *Deleuze and Space*, 36–61 (Edinburgh: Edinburgh University Press).

Helgadottir, G. (2008), *Textiles of the North*, document by personal communication, *in preparation.*

Jewesbury, D. (2003), 'Tourist:Pioneer:Hybrid: London Bridge, the Mirage in the Arizona Desert', in Crouch, D. and N. Lübbren (eds) *Visual Culture and Tourism* (London: Berg).

Jones, N. (1997), 'The Perception of Character in the Phenomenology of Paintings, in Context and Value: Art History and Aesthetic Judgments', unpublished Masters thesis, University of Warwick.

Kirshenblatt-Gimblett, B. (1998), *Destination Cultures: Tourism, Museums and Heritage* (Berkeley, CA: University of California Press).

Massey, D. (2005), *For Space* (London: Sage Publications).
Matless, D. (1992), 'An Occasion for Geography: Landscape, Representation and Foucault's Corpus', *Environment and Planning D: Society and Space* 10:1, 41–56.
Miller, D. (1987), *Material Culture and Mass Consumption* (Oxford: Blackwell).
Miller, D. (1997), 'Coca-Cola: A Black Sweet Drink from Trinidad', in D. Miller (ed.) *Material Cultures*, 169–87 (London: UCL Press/University of Chicago Press).
Miller, D. (ed.) (1998), *Material Culture: Why Some Things Matter* (London: Routledge).
Moore, N. and Whelan, Y. (eds) (2007), *Heritage, Memory and the Politics of Identity: New Perspectives on the Cultural Landscape* (Aldershot: Ashgate).
Pearce, L. (2004), *Driving North/Driving South: Reflections Upon the Spatial/Temporal Co-ordinates of Home*, Discussion Paper, Lancaster Mobilities Group Lancaster University.
Rose, G. (1993), *Feminism and Geography* (Minnesota: Minnesota University Press).
Rose, G. (2001), *Visual Methodologies* (London: Routledge).
Taussig, M. (1992), *The Nervous System* (London: Routledge).
Thrift, N. (2008), *Non-Representational Theory* (London: Routledge).
Tolia-Kelly, D. (2008a), 'Fear in Paradise: The Affective Registers of the English Lake District Landscape, Re-Visited', *Senses and Society* 2:3, 329–51.
Tolia-Kelly, D. (2008b), 'Motion/Emotion: Picturing Translocal Landscapes in the Nurturing Ecologies Research Project', *Mobilities* 3:1, 117–40.
Urry, J. (2002), *Sociology Beyond Societies: Mobilities for the Twenty-first Century* (London: Sage).
Urry, J. (2003), *The Tourist Gaze* (London: Sage Publications).
White, T-A. (2004), 'Theodore and Brina: An Exploration of the Myths and Secrets of Family Life, 1851–1998', *Journal of Historical Geography* 30, 520–30.
Williams, R. (1989), *Resources of Hope: Culture, Democracy, Socialism* (London: Verso).
Winter, T. (2007), 'Landscapes in the Living Memory', in N. Moore and Whelan, Y. (eds) *Heritage, Memory and the Politics of Identity: New Perspectives on the Cultural Landscape*, 133–48 (Aldershot: Ashgate).

PART II
Representation and Substitution

Chapter 5
The Popular Memory of the Western Front: Archaeology and European Heritage

Ross Wilson

The Western Front retains an important place in the 'collective memory' of the former combatant countries of Europe, Canada, Australia and New Zealand. The conflict still lingers in the minds of many, even as the 100th anniversaries of the war approach. In many respects, however, the history and memory of the conflict remain constricted by national agendas. In Britain, the battlefields of Northern France and Belgium appear to still haunt contemporary society, as despite fierce fighting in other theatres of war, the Western Front is remembered as *the* Great War (Bond 2002). Indeed, such is the effect of this myopia that the war on the Western Front is often viewed as a 'private British sorrow' (Terraine 1980, 120). The poignant hold of the battlefields and the resonance that 'the trenches' still possess in Britain can be witnessed in the novels and films still produced every year (Korte 2001, 120). The images of desolate war-torn landscapes, shattered tree stumps and disillusioned soldiers occupying filthy, muddy, rat-infested trenches still possess the capacity to evoke deep emotion nearly 90 years after the Armistice. Alongside these disturbing images, the conflict is recalled with images of apparently peaceful serenity, of the cemeteries of the Commonwealth War Graves Commission, where the monumental architecture of Luytens and Blomfield, the uniform headstones and verdant turf were utilized in the immediate post-war period to 'heal a scarred landscape' (Heffernan 1995, 293). It is these contrasting images of the battlefields of Northern France and Belgium that serve as the visual frames of reference for the memory of the Western Front.

Despite the evident differences between these two sources, both sets of images have contributed to the popular memory of the war in Britain. This popular memory of the soldier of the Great War is one coloured by pity and poignancy for the men who fought; remembering the passive soldier as a victim of the war, fighting in atrocious conditions, ordered by an indifferent officer class. This perception has been challenged in recent years by historians wishing to revise this popular memory, who seek to assert the achievements of the British Army on the Western Front (Griffith 1994). Over the last decade, a further source for the memory of the battlefields has been provided by the advent of archaeological investigations on the former fields of conflict. Excavations of material from the trenches and no man's land of the world's first industrialized war, and the remains of the men who fought, provide a distinct contribution to the memory of the conflict. As yet, the impact that archaeology will have on the way in which the war is remembered has yet to be assessed. As excavations in the region increase, however, attracting growing amounts of interest in the media and amongst the general public, such an assessment is essential. It is the images from these excavations, which are perhaps the most distinctive and powerful part of archaeology's recent involvement in the battlefields, that will be at the core of this study, as it is taken that the image, however constructed, is central in the formation of the memory of the recent historical past (Hall 1991, 153). This study therefore constitutes an examination of the 'image in memory'; how the visual representation of the past is a vehicle for the expression of popular memory. Through the analysis of the visual representation of the Western Front within Britain, the use and value of the image in remembering the past will be explored.

The Image in Memory

The role of the image in the formation and retention of memory has a long history. Yates (1966) has highlighted how the mnemonic devices of the classical world depended upon the image, but also how the image facilitated a form of 'sensuous memory.' Indeed, within Greek and Roman culture the 'process in which impressions were received and retained by the memory was generally believed to be somatic or bodily' (Whitehead 2003, 12). One of the most well known of these mnemonic practices was that supposedly used by Simonides, the Greek poet of Cos in the fifth century. Simonides was able to recall those who were present at a banquet, where the roof of the hall had collapsed crushing those inside, ensuring that their bodies were unidentifiable. By attaching his memory to the image of the banquet and immersing himself into that image, the poet was able to recall who had been present at the feast (Yates 1966, 1–2). For the development of this technique, therefore, memory came to be considered as a corporeal experience which centred on the image. As Aristotle (1972, 62) stated, 'memory, even the memory of objects of thought, is not without an image'. Memory was considered as an imprint of an image on the mind. Plato (1973, 191) compared memory to an image on a wax tablet, stating that:

> If there is anything we want to remember … we hold it under the perceptions and conceptions and imprint them on it, as if we were taking impression of signet rings. Whatever is imprinted we remember and know … but whatever is smudged out or proves unable to be imprinted, we've forgotten and don't know.

It is indeed this link between memory and image that has provided one of the most enduring ways in which memory is conceived, structuring the way we view and conceptualize how memory works.

Mnemonic practices as formulated in the ancient world were therefore a pictorial art, focussing not on words but on images to create a scene or place within which the individual could place the images that were to be recalled. This use of memory came to be classed as a branch of rhetoric (Yates 1966, 10). The mnemonic techniques of classical Rome, for instance, were developed in reference to oratory, whereby a Roman lawyer could use the classical technique of fixing memory to an image as a way of remembering meaningful points of an argument he wished to place before the court. The orator in these circumstances would imagine walking through a building and attaching parts of the speech to the places and objects in that building to sustain memory (Fentress and Wickham 1992, 12). As the first century AD Roman orator Quintilian (2001, 76) stated:

> Students learn sites (*loca*) which are as extensive as possible and are marked by a variety of objects, perhaps a large house divided into many separate areas. They carefully fix in their mind everything there which is notable, so that their thoughts can run over all the parts of it without any hesitation or delay.

Saint Augustine and Saint Thomas Aquinas also utilized the image and this concept of memory in their writings on the scriptures. According to Saint Thomas Aquinas, who based his work on memory on the teachings of Aristotle, the art of connecting images with memory was the very essence of remembering: 'Man cannot understand without images' (Yates 1966, 70). This art of memory allowed the individual to become physically immersed in their theological teachings, so that they were able to interpret the world around them through their understanding. This articulation of the image in memory was also apparent in Renaissance Europe. The rediscovery of the works of Plato and Aristotle by scholars, and traditions of neo-platonic thought, led to a reconsideration of this sensuous memory to aid recollection, with the construction of 'memory theatres' (Yates 1966, 148). The corporeal experience of memory in these 'memory theatres' acted upon the individual to recall past experiences in a manner which enabled the mobilization of emotion and imagination, calling into performance body and soul.

Memory in the Mind and the Body

The power of the image to impact upon and assist memory has persisted to the present day, though the regard for the sensuous experiences of memory has not survived. Following the Enlightenment, philosophers such as David Hume and Thomas Reid, who pioneered the first positivist analysis of memory, the ability to recall began to be reconsidered solely as a cognitive function. This definition of memory has continued throughout the twentieth century, with the psychological studies of memory confirming memory's role as a mental facet (Freud 1957). More recently, the relationship between memory and the image has been debated within the context of mass production and the advent of the post-modern critique. In this regard, it has been argued that the image, whether as a photograph or transmitted through the mass media, aids the derogation of modern memory.

In a society dominated by constantly fluctuating and potentially manipulating images, what can be termed 'real' or 'organic' memory can be seen to have diminished, such that, as Baudrillard has noted, 'the real has become the hyperreal' (Baudrillard 1988, 5). In this manner, photographs were viewed, 'not so much an instrument of memory as an invention of it or a replacement' (Sontag 1990, 165). Images of the past are articulated in this respect out of their original contexts and utilized for the purposes of the mass market. In response to this criticism of late capitalism hijacking the image, Berger (1997), in his discussion of photography and memory, has drawn a distinction between the public and the private image. The public photograph in this perspective, used, for example, by corporations for marketing objectives, is considered to have 'been severed from life when it was taken, and it remains, as an isolated image, separate from your experience. The public photograph is like the memory of a total stranger' (Berger 1997, 44). By contrast, the private image has retained its meaning by becoming infused with social memory; it is attributed with significance and value by those who view it.

The image, in this respect, does not have to be of an event experienced in person, or of an individual personally known to the observer, but one to which importance is attached by the observer. This is the *punctum* as described by Barthes (1984, 198), the point of entry into an image by the observer. In this respect, the photographic images of a recent historic past, though still distanced from contemporary society, provide a means to access, experience and remember. Rather than contributing to the pastiche of late capitalist society, or demonstrating the absence of 'the real', images of the past in this respect offer an opportunity to explore the past, and examine concepts of 'place' in a wider cultural context (Berger 1997, 47). Following this perspective, Landsberg's (2004) 'prosthetic memory' and Bennett's (2005) 'empathetic memory' have described how the image can provide an original perspective in remembering a historic past. Essentially, both approaches have assessed the accessibility and immediacy of the image in memory, enabling the present observer to witness and possess a memory of the past that they did not directly experience. This feature of the image of the past allows the viewer to feel a connection to a past that they have witnessed through the image, a sensuous experience for the viewer as they make a 'leap into the being of the past' (Deleuze 1973, 65). This, in effect, represents a phenomenology of memory, whereby the corporeal experiences of the scene or of the individual depicted in the image is communicated to the observer. It is through this phenomenology of memory that the archaeological images of the Western Front will be explored.

The Memory of the Western Front

The names of the battlefields of Northern France and Belgium still evoke powerful emotions in Britain. That the Western Front still holds sway over British popular memory is in large part due to three factors. Firstly, the majority of British troops serving in World War One fought and died in this theatre of war. Secondly, Britain had never before witnessed such a mobilization of civilians for the conflict. Thirdly, the vast numbers of injuries and deaths on the battlefields of Northern France and Belgium have ensured that this front was, and still is, the bloodiest in British military history (Corrigan 2003). In the context of the twentieth century it is also inevitably remembered in contrast to the World War Two. Whilst the 1939–45 conflict is regarded as a 'just war', a morally and politically correct conflict, the Western Front is viewed as a meaningless disaster. The Western Front is recalled in Britain through the image of 'the trenches': mud, blood, rats, gas, incompetent generals and the 'lost generation'. The battlefields are recalled through the poetry of Wilfred Owen and Siegfried Sassoon, especially the former with his stunning indictment of the 'old lie' of *Dulce et Decorum Est*. The trenches are recalled as places of infinite horror punctuated only by Christmas Truces. They are remembered through popular television programmes and films, *Blackadder Goes Forth* and *Oh! What a Lovely War*. Overwhelmingly, the Western Front in Britain is remembered as a tragic, futile waste. This perception is reinforced by the words and phrases associated with the conflict. Whilst these are used within the context of daily life, they gain colour and added significance through their associations with the battlefields. Terms such as 'going over the top', 'entrenched', or 'in the trenches' rely upon a popular cultural memory of the Western Front as one of cruel, industrialized war and noble, suffering soldiers. Military historians have argued that this perception of the battlefields is one which is entirely based on myth. These studies have asserted that *the trenches* as the scene of unimaginable misery has been created and sustained through misrepresentation by the war poets and by successive depictions in literature, film and television (Bond 2002).

To act as a rebuttal towards this perceived dominance of cultural memory over historical rigour, a number of military historians have moved to dispute the basis of the belief that the Western Front was a national tragedy. Through the analysis of the British Army during the conflict, its advances in arms, tactics and logistics, not to mention, of course, that the Allied Forces did succeed in securing victory in 1918, these scholars have attempted to shift the Western Front from memory to history. The 'myth' of the war poets who stressed disillusion and horror in the trenches are countered here with the testimonies of ordinary veterans who describe admiration for generals, respect for comrades and positive wartime experiences (Sheffield 2002). The popular memory of the Western Front is also deconstructed by scholars keen to emphasize that the image of 'mud and blood' that the battlefields appear to conjure almost automatically deters serious and sober study. Bond (2002, 5) has tracked the chronological development of the popular memory of the battlefields and emphasized the formative role media representations have played.

The publication of certain novels, the release of particular films and the broadcast of television programmes are regarded here as the origin of the memory of the battlefields (Todman 2005). This is evidenced by the recent campaign to issue pardons to British soldiers executed by the army during wartime, a movement which sought to emphasize that they were also 'victims of the war'. The rise of the notion of victimhood to describe the soldiers of the conflict is regarded as a contemporary phenomenon by military historians. This victim-status was most prominent during the 80th anniversary of the Armistice in 1998 and followed on from the 'mini-boom' in war books in the early 1990s (Korte 2001, 121). The popular novels of Pat Barker's trilogy, which began with *Regeneration* (1991) and Sebastian Faulk's *Birdsong* (1993), are viewed as reflecting and encouraging this trend. However, such is the powerful and emotional grip that the memory of the Western Front induces; the efforts at revisionism have been largely unsuccessful.

Following on from this deconstruction of the popular memory, it is worthy of note to identify that the Western Front is not wholly definable or describable in myth, memory or history. The 'truth' of the war does not exist solely for military historians to assert. All forms of memory and history are created through various sources. Interpretations of the conflict from the perspective of military history are of course organized, selected and filtered just as any other interpretation. What has occurred since the cessation of the conflict is a reoccurring process of reinterpretation. The representation of the conflict, which is formed through military history, has been supplemented and, on occasion, replaced by other forms of historical discourse; oral history, family history, cultural history, social history and micro-history. These have inevitably drawn on differing sources of evidence, theories and methodologies to build alternative representations. In this manner the Western Front forms a 'site of memory' that is being continually negotiated, retold and remembered from a myriad of viewpoints (Korte 2001, 121). This alteration and reformation of memory is demonstrative of distinct cultural, societal and generational differences. It is also evidence of the fluctuating drives, desires and requirements of those remembering the battlefields. In this manner, the rearticulation of the memory of the Western Front is essential entirely to perpetuate 'the memory' of the battlefields. Studies of the memory of the wars and conflicts of the twentieth century, an ever-growing field, have frequently reiterated this aspect of renegotiation in remembrance (Ashplant et al. 2000). This argues against Nora (1989), who would view these renegotiations of memory as desperate attempts to gather and prolong that which we can no longer possess.

Historians, therefore, in the quest to 'debunk' the myths and cultural memory of the war by locating their source on the novels, film and television representation of the Western Front, have neglected the study of the formation of popular memory. The implicit conjecture behind these statements is that these images of the past are chosen by the public because they are 'easier'. The transmission of the past through the mass-media is associated here with a rather vapid consumerism. This analysis appears insufficient when one considers the way in which the Western Front appears to haunt the popular consciousness; providing the archetypal image of industrialized war and meaningless slaughter. We can therefore assess the cultural memory of the battlefields as one which is preferred over others. This is to acknowledge that cultural memory carries with it value and purpose for the population that honours it. Samuel and Thompson (1989, xxviii) wrote that, 'In dealing with the figures of national myth, one is confronted not by realities which became fictions, but rather by fictions which by dint of their popularity, became realities in their own right.' The revisionist histories of the Western Front that celebrate the achievements of the British Army are therefore compared with how a wider public uses this history and selects this history, constructing different stories and narratives of the past. This construction of cultural memory has too often been regarded by historians as a barrier to 'objective' study. The cultural memory of the battlefields has been undermined in this respect without due consideration of the 'selection, ordering and simplifying' which occurs within cultural memory (see Samuel and Thompson 1991, 18). Whilst historians have spent a great deal of energy examining the genealogy of the cultural memory of the battlefields, the motivations for the veneration of this memory have been neglected. The memory of the Western Front in Britain should therefore be regarded as specifically chosen rather than unthinkingly grasped.

The Image of the Western Front and British Nationalism

Through the analysis of the cultural memory of the Western Front in Britain as a 'preferred choice', wider notions of memory, identity, place and belonging can also be explored. By referring back to the recent popular novels that represent the battlefields and are criticized by revisionist historians, the Western Front can be examined as a vehicle for articulating a sense of self. This feature of the cultural memory of the battlefields is most evident in the upsurge of recognition of 'suffering' in the trenches. The British soldiers as tormented 'victims' of the war is an especially interesting feature of these novels. Barker (1991, 45) in fact sets the first book in the *Regeneration* trilogy at the Craiglockhart Hospital where soldiers received treatment for a variety of maladies caused by the war, but especially for the condition of 'shellshock'. It was at the Craiglockhart Hospital, where the pioneering work of Dr W.H.R. Rivers identified and treated this war-induced trauma. The use of the condition by Barker in the novel is, however, employed to emphasize and reiterate the distress and anguish of men. Shellshock forms a means of highlighting the victim-status of the soldier. Rather than examine the role of treatment, it builds a new means of emphasizing the tragedy of war (Korte 2001, 124). Barker's use of shellshock in this way should be considered in the societal context to examine the uses of the image of 'soldiers as victims'.

The condition of shellshock gained wider public prominence throughout the 1990s as popular and academic histories addressed the issue in the context for the campaign to pardon soldiers executed by the British Army during wartime. Over 300 soldiers were executed during the conflict for a variety of reasons. However, many commentators asserted that many soldiers who were executed for cowardice or desertion could well have been suffering from shellshock. This cause received such vocal public support that although the appeal for pardons was denied in 1998 it was eventually granted in 2006. Nevertheless, during the 80th anniversaries of the Armistice in 1998, Government statements also described executed soldiers as 'victims of the war'. Shellshock, trauma and victimhood are indicative terms that mark the forms of remembrance in the late 1990s as distinct. The 80th anniversary of 1918 can be observed as a moment when the pain and trauma of the conflict was rearticulated by the wider population. This marked an occasion when popular sentiment demanded that the soldiers of the Western Front be remembered as victims of the war (Moriarty 1999).

This form of remembrance cannot be proscribed to a previous absence of history or literature regarding the perspective of the ordinary soldier. The advent of oral histories in the 1960s ensured that 'the view from the trenches' had long been recognized. What marked the 80th anniversary was a new discourse of remembrance associated with the notion of victimhood. Novick (1999, 5), through his work on the Holocaust remembrance in America, has suggested that this association with suffering reflects the wider cultural shifts that occurred in the late twentieth century, which has placed the notion of 'passive suffering' and 'vicarious victim-hood' as a desirable, and even an advantageous, quality. In this way, to state links to those in the past who have endured pain and trauma is a significant tactic in the demand for recognition and the validation of identities. If the formation of identity at any level is regarded as a continuous cycle of discourse that draws upon history and memory to represent and confirm present struggles, then the memory of the Western Front can be seen as a significant site of remembering.

In this respect, the wider societal context of the late 1990s provides the setting in which to examine the emergence of 'victim-hood'. As Scotland and Wales in 1997 and Northern Ireland in 1998 achieved the promise of some form of independence, and with the increased debates regarding wider European integration, the concept of the British soldier suffering in the trenches can be seen as a means for sections of the population to coalesce around a single issue. This movement to affirm a particular memory of the past appears significant as the hereto well-established national boundaries appeared to be crumbling. The marking of the 80th anniversary of the cessation of hostilities in 1998 can only be described as cathartic. The event witnessed a public voicing of pain and anguish regarding the soldiers of the world's first industrialized war. The issues which arose from this, of victim-hood, trauma and shellshock can be interpreted along the lines of establishing and expressing a national identity. As the popular historian of the war Lyn MacDonald stated at the time, 'people are interested because they *care*, people *care* about this ... People realize it is relevant to our country now' (quoted in Stummer 1998, 23).

The image of the Western Front therefore informs and shapes concepts of British nationalism; the representation of the British soldier who endures suffering and trauma in the trenches appears to reinforce the national, 'British' virtues of stoicism, honour and endeavour. But this connection and association are not inevitable. Indeed, there are a variety of ways in which images of the Western Front can be read and the battlefields thereby remembered. Re-imagining the image of the battlefields can provide alternative memories and the role of the archaeological image is vital in this process (Russell and Cochrane 2007). It is through the archaeological image of the Western Front that an alternative remembrance can be offered of the former battlefields. This remembrance uses the popular memory of the Western Front, but expands its scope beyond its British focus, to consider the existence, the corporeal being of 'an other' and an other's history and memory. This considers the strong connection between image and memory, and the formation of memories of the historic past through the image.

The Image of the Western Front

It is this capacity to impact upon memory that the images of the excavations of the Western Front can best be understood. The pictures of excavated trenches still lined by rusted corrugated iron, with the trench boards visible through the mud of Flanders and Northern France, and photographs of human remains, with the remnants of boots and uniforms still observable, has provided haunting and evocative images that have been published widely in newspaper reports and broadcast on television. It is these images that have had the greatest impact around the world, especially in terms in legitimizing and popularizing an archaeological agenda on the battlefields. The recent excavations near the Belgian town of Ieper in advance of the construction of the A19 motorway indicate how these images serve to 'pluck the chords of memory' (Denning 1994, 353). On 11 November 2003, reports of excavations that had unearthed sections of the front line held by British, as well as Allied and German soldiers, discovering human remains, the material of the war, and the trenches and dugouts of the battlefields emerged in various news bulletins in Britain. The added poignancy of choosing Remembrance Day on which to report this excavation, which had been already ongoing for several months, caused a widespread public response regarding the responsibility and necessity to maintain the memory of the Great War on the Western Front (Black 2003).

Though the inception of archaeology on the Western Front occurred over a decade ago, the ways in which it can shape and influence the popular memory of the battlefields has yet to be fully examined. The potential for the archaeological analysis of the Western Front is significant as the material remains of the four years of war are substantial. As Saunders (2001, 38) has observed, excavations uncover, 'trenches, dugouts, material and human remains often perfectly preserved, just centimetres beneath the surface.' The objects associated with the lives and deaths of the soldiers who fought here still litter the former fields of conflict. Whilst the use of archaeology in the remembrance of the conflict is limited, a significant degree of attention from the public and the media has been drawn to the remains of soldiers that are frequently unearthed. The controversy that was aroused by the broadcast by a television programme documenting the activities of the Belgian amateur archaeological group, 'The Diggers', as they excavated the remains of a British soldier is testament to this (Barton 2003). Saunders (2001, 49) has even compared the sensitivities involved within these situations to the ongoing debates regarding the repatriation of Indigenous artefacts and human remains from Western museums to their descendant groups. This comparison is especially pertinent when one considers the excavation and repatriation of 'Unknown Soldiers' by Australia in 1993, Canada in 2000 and New Zealand in 2004 to form their own national memorials. As the controversies regarding the excavation of human remains from the battlefields continued, a number of archaeologists continued to excavate the trenches and front lines of the conflict. These excavations revealed issues which work with and against the popular memory of the conflict in Britain, as they concerned the violence, brutality and death in the landscape (Desfossés et al. 2003).

One of the first excavations of the battlefields of the Western Front is perhaps the most illustrative of the way in which archaeology and the archaeological image informs cultural memory. In 1991 at Saint-Rémy-la-Calonne near Verdun in Northern France, a mass grave of French soldiers was excavated (Hill and Cowley 1993). The site was located and excavated largely for its association with the writer Henri-Alban Fournier, known as Alain-Fournier, author of *Les Grand Meaulnes*. Fournier was part of the 288th French Infantry Regiment, which disappeared during an attack on a German unit in September 1914. Field surveys revealed the location of the grave and permission was sought from the authorities to excavate. The grave measured approximately 5m by 2m and contained the remains of 21 individuals. The excavation revealed the bodies had been carefully laid out at the time of burial, in two rows of ten, head-to-foot, with the last body placed horizontally across the others. The subsequent material and document analysis also indicated that the men were buried according to their military hierarchy. The Captain of the regiment was laid out in the grave first at one end, followed by the Lieutenant, the Second Lieutenant, followed by the other ranks. This apparently sympathetic burial of the soldiers was in marked contrast to the popular memory within France that Fournier and his men were executed by their enemies. Analysis of the remains showed that the men had died from wounds sustained in combat. Six men did, however, have single bullet wounds to the head, although these may have been mercy shots, as their skeletons showed signs of sustaining other injuries. Images of the mass grave and its excavation were widely distributed across France enabling another vision of the war to emerge (Hill and Cowley 1993, 430).

Further excavations have revealed the nature of violence and brutality on the battlefields but also how soldiers lived and died. They have provided striking images of the human experience of conflict. For instance, at Gavrelle, Pas-de-Calais, archaeological excavations unearthed a mass grave with the remains of 12 German soldiers. Examination of the material remains indicated that most of those buried had had their personal and military effects taken from them before burial. Significantly, the discovery of small eyelets at the top of the excavation indicated that some men had been interred wrapped in canvas tents. The presence of regimental insignia found on the remains of the soldiers' uniforms enabled identification of the soldiers as belonging to the 152nd Infantry Regiment of the 48th Division of the German Army. Troops belonging to this regiment were known to have sustained fatalities during the German Spring Offensive of 1918. This historical research, coupled with their apparently hasty burial, suggested that they were buried quickly by retreating Allied soldiers (Desfossés et al. 2000). This quick burial in the battle zone is in contrast to a burial of a British soldier located close by at Monchy-le-Preux. Here, the burial of the soldier appears to have been made with care and attention, as the body was carefully laid out in the fill of an old shell hole. This is in contrast to the remains of a German soldier discovered in the same shell hole. By the position of the skeleton the soldier appears to have died where he fell and subsequently buried by the falling debris from a shell blast (Desfossés et al. 2000).

The excavation in 2001 of a mass grave at Le Pont du Jour, near Arras in northern France, reinforces the archaeological study of the battlefields. It also provides an insight into the value of the archaeological image and its role in the formation of memory. In this excavation, the bodies of 20 British soldiers were found carefully laid-out, presumably by their comrades. An upturned shell case indicates that the grave was probably marked in the expectation that the bodies would be recovered and re-interred by a burial party. The extensive physical trauma of the war was especially in evidence on the skeletons. One of the soldiers in the burial was found to have been hit by three pieces of shrapnel, one piece cutting his larynx, the second entering his rib cage and possibly perforating his heart, and the last cutting into his right knee (Bura 2003, 94). This excavation in particular drew attention for the first time to the way in which soldiers buried their comrades. It is this aspect which also served to act upon the popular memory of the conflict in Britain. The haunting images of a line of skeletons placed side by side by those who served alongside them fuelled speculation in the national press that it demonstrated the deep and affectionate bonds created during wartime. This seemed to link in with the memory of the noble, heroic soldier fighting in brutal conditions, but maintaining a sense of camaraderie.

Archaeology, the Image and a European Heritage

The role of images in archaeological narratives is a subject only recently addressed within post-processual archaeology (Moser and Smiles 2005). Images within archaeology have often been considered as merely technical aids, as diagrams or as 'objective' recordings of features and artefacts, rather than essential parts of the narrative, as another voice in the stories which archaeologists tell through their interpretations, or a spur to the remembrance of the past. Rather than allow archaeologists to rely on a conception of image use as a functionary means of displaying data, post-processual archaeologists have used images to challenge and subvert interpretations of the past within the narratives they construct. Archaeological images have undoubtedly a significant impact on the ideas generated about the past and about archaeology itself. Therefore, these images also contribute to the memory of the past in the present, as Bertrand Russell (1921, 75) stated, memory can 'only be represented by images.' It is in this context that the images of the Western Front can be considered, as whilst these photographs and images of archaeological contexts do not provide 'innocent analogues' of the past, they do represent a means of exploring the embodiment of the materials and spaces of the past, and form an important and valuable research tool (Shanks 1997, 101). The images from archaeological investigations of the spaces of the war, and the material and human remains, also offer a means by which a new narrative of the war on the Western Front can be told.

A war which ended nearly 90 years ago on the Western Front still resonates today, not just in Britain but across Europe. This was also a war which witnessed European nations engaging in conflict for national and imperial aims. As European initiatives have searched for historical periods whereby European integration or a specific European character could be fostered and legitimated, the Western Front could become a key feature in this process. Archaeologists have already become involved in this process, but as yet these projects have concentrated on prehistoric examples, away from the far more apparently troublesome European history of the twentieth century (Renfrew 1994). The memory of the world wars fought in Europe is even perceived by some to be the most difficult obstacle to overcome if a move towards European integration is to be realized. Smith (1995, 133) has stated that the European memory is 'haunted by war', and that 'if its peoples share only the painful reminders of a nationally divided past, can they perhaps unite around common myths and symbols which signal a deeper solidarity and difference?' This is perhaps especially so in Britain, where the popular memory of the conflicts is constantly reiterated and serves to stress the distinctions between Britain and the rest of Europe, and casts European nations as 'the other.'

The popular memory of the Western Front in Britain certainly contributes to this perspective, as the battlefields of Northern France and Belgium are remembered in an insular, national fashion, as a defining moment in British national history. These sites of war, however, which were once scenes of bitter national divisions, can conversely act to unite nations individually and collectively in mourning and remembrance in alternative narratives of the war. The images of the archaeological excavations can play a vital role in this retelling as, by utilizing Landsberg's (2004, 8) approach to the way in which images can create a sensuous memory of a past not directly experienced by the observer, it opens up the memory of the Western Front. It facilitates remembering in a manner which allows a greater contemplation of the experiences of the British soldiers, and allows others to view and share these experiences as well (Levinas 2001, 100). It is this exposure in memory, the regard for expressing the corporeal experiences of the past, which allows this memory to be considered by 'the self' and the 'other'. Allowing 'the other' to view what may be considered by some as the treasured, national historical narratives in this manner is a major factor in undermining the insular nature of the memory of these events. Indeed, the manner of such a disclosure can be considered as an antithesis to the way in which the singular national memory of the past fosters distrust and suspicion between peoples. Indeed, as Wittgenstein (1998, 52e) argues, 'hate between human beings carries from our cutting ourselves off from each other. Because we don't want anyone else to see inside us, since it's not a pretty sight in there'. By letting someone see inside us, by remembering the past in this particular mode, we can avoid becoming both strangers to ourselves and 'our' own history and to others by hiding and obscuring 'our' past.

It is also through this approach to the image in memory, by looking at the images of excavated trenches and human remains from all combatant countries, that a further possible agenda for considering the battlefields in the context of 'European heritage' can be observed. It is this regard for the corporeal experiences in memory that allows for a consideration of Levinas's (2001, 21) concept of responsibility. Focusing on the image in this manner enables an 'interhuman' perspective; both through a consideration of how all soldiers on the front lived and died, but also through acknowledging another's history and memory of the past. Levinas (2001, 100) has argued that the possibility of finding meaning within our own cultural memory can only be borne out by a consideration of the presence of an others' memory of their historical past, their pain, their suffering, their pleasure, their existence. The images from archaeological projects on these former battlefields can also play a vital role in this retelling. The pictures of the excavation of the remains of soldiers in particular provide an opportunity to reflect on the past, and its relevance to the present. The photographs of the mass graves at Saint-Rémy, Le Pont-du-Jour and Gavrelle, although representing the dead from different nations, are striking in their visual similarities when compared to the nationalistic and patriotic designs of the national war cemeteries. These images provide a means for recognition and consideration of the pain and experience of others, and perhaps can even form the basis of a shared *momento mori* in Europe. As Sontag (2003, 102) states,

> … let the atrocious images haunt us. Even if they are tokens and cannot possibly encompass most of the reality to which they refer, they still perform a vital function. The images say: This is what human beings are capable of doing – may volunteer to do, self-righteously. Don't forget.

Conclusion

The image of the Western Front as a senseless waste, a poignant national tragedy for Britain, has been prominent in the popular memory of the conflict. This memory of the battlefields has been chosen from the array of representations in the media and elsewhere as expressing the 'truth' of the war. As archaeological excavations continue to produce startling reminders of the conflict, this remembrance can begin to take alternative forms. This is not to assume that archaeology has a greater prominence than other modes of remembering the Western Front, whether through historiography, the memorial landscape, literature, film or television; rather, an understanding of the role of the archaeological image in this nexus is required (Wilson 2007, 227). By placing the archaeological image in the context of the popular memory of the suffering soldier on the Western Front in Britain, a comprehension of the conflict as 'European heritage' can be considered. This European heritage, rather than being based on inevitably exclusionary concepts of shared identity or beliefs – a divisive issue itself in the multicultural European Union – is rather based on a shared set of responsibilities towards 'the other' (Derrida 1992, 79).

The tendency of the Western Front to accentuate aspects of British nationalism is not a certainty in any respect; the memory of the Western Front is in a constant state of flux. By realizing that the popular memory of the trenches of the Western Front is specifically chosen by its audience, that the trenches form a symbolic resource for society to draw upon to express concerns and fears, and that this feature is itself valuable and worthy of study, a greater regard for the popular memory of the Western Front can be formed. Too often, critics have lamented the apparently easier representations chosen by the public, citing television series such as *Blackadder Goes Forth*, novels such as Barker's *Regeneration* trilogy, and films such as *Oh! What a Lovely War*, as providing inaccurate impressions of the battlefields which are vapidly consumed by a credulous public. To reject this position enables a perspective that can begin to assess the motivations behind the articulations of the popular memory of the trenches, and value them as significant contributions to the study of the Great War and society. This is of particular contemporary concern with regard to an international memory of the battlefields, as increasingly the popular memory of the trenches is invoked as a means of criticizing the recent war in Iraq. The revivals of *Journey's End* and *Oh! What a Lovely War* on the West End and touring productions around Britain, as well as the re-release of the film *All Quiet on the Western Front* in the last three years, have all played against the backdrop of the Iraq War, allowing the audiences to draw parallels with the images of a slaughter of a war of attrition, the tragic farce of a war without meaning, and a critique of a government's policies that instigated the conflict. Critics of the popular memory have underestimated the value of the memory of the trenches of the Western Front in the representation of current concerns; failing to recognize that to an extent every generation will return to the trenches as it seeks to utilize a powerful memory as a vehicle to express itself.

References

Aristotle (1972), *Aristotle on Memory* (London: Duckworth).

Ashplant, T., Dawson, G. and Roper, M. (2000), 'The Politics of War Memory and Commemoration: Contexts, Structures and Dynamics', in T.G. Ashplant, G. Dawson and M. Roper (eds) *The Politics of War and Commemoration*, 1–78. (London and New York: Routledge).

Barker, P. (1991), *Regeneration* (London: Penguin).

Barthes, R. (1984), *Camera Lucida* (London: Fontana).

Barton, P. (2003), 'The Corner of a Foreign Field that will No Longer be Forever England', *Battlefields Review*, 23, 13–5.

Baudrillard, J. (1988), *Selected Writings* (Cambridge: Polity Press).

Bennett, J. (2005), *Empathic Vision: Affect, Trauma, and Contemporary Art* (Stanford: Stanford University Press).

Berger, J. (1997), 'Ways of Remembering', in J. Evans (ed.) *The Camerawork Essays: Context and Meaning in Photography*, 41–51. (London: Rivers Oram Press).

Black, I. (2003), 'Row Over Fate of First World War Trench Unearthed on Belgian Motorway Route', *The Guardian* [Online, 11 November]. Available at: http://www.guardian.co.uk/international/story/0,,1082389,00.html [accessed: 11 November 2005].

Bond, B. (2002), *The Unquiet Western Front* (Cambridge: Cambridge University Press).

Bura, P. (2003), 'Étude Anthropologique de la Sépulture Multiple', *Sucellus*, 54, 92–8.

Corrigan, G. (2003), *Mud, Blood and Poppycock: Britain and the Great War* (London: Cassell).

Deleuze, G. (1973), *Proust and Signs* (London: Allen Lane).

Denning, G. (1994), *Mr. Bligh's Bad Language: Passion, Power and Theatre on the Bounty* (Cambridge: Cambridge University Press).

Derrida, J. (1992), *The Other Heading: Reflections on Today's Europe* (Bloomington: Indiana University Press).

Desfossés, Y., Jacques, A. and Prilaux, G. (2003), 'Arras "Actiparc" les Oubiliés du "Pont du Jour"', *Sucellus*, 54, 84–91.

Faulks, S. (1993), *Birdsong* (London: Harper Collins).

Fentress, J. and Wickham, C. (1992), *Social Memory* (Oxford: Blackwell).

Freud, S. (1957), *Civilisation and its Discontents* (London: The Hogarth Press).

Griffith, P. (1994), *Battle Tactics of the Western Front: The British Army's Art of Attack 1916–1918* (New Haven: Yale University Press).

Hall, S. (1991), 'Reconstruction Work: Images of Post-war Black Settlement', in J. Spencer and P. Holland (eds) *Family Snaps: The Meaning of Domestic Photography*, 152–64. (London: Virago).

Heffernan, M. (1995), 'For Ever England: The Western Front and Politics of Remembrance in Britain', *Ecumene*, 2:3, 293–323.

Hill, D. and Cowley, D. (1993), 'A Bundle of Presumptions: Military Archaeology Solves a Literary Mystery', *Quarterly Journal of Military History*, 6:1, 410–34.

Korte, B. (2001), 'The Grandfathers' War: Re-imagining World War One in British Novels and Films of the 1990s', in D. Cartmell, I.Q. Hunter and I. Whelehan (eds) *Retrovisions: Reinventing the Past in Film and Fiction*, 120–31. (London: Pluto Press).

Landsberg, A. (2004), *Prosthetic Memory: The Transformation of American Remembrance in the Age of Mass Culture* (New York: Columbia University Press).

Levinas, E. (2001), *Ethics and Infinity: Conversations with Phillipe Nemo* (Pittsburgh: Duquesue University Press).

Moriarty, C. (1999), 'Review Article: The Material Culture of Great War Remembrance', *Journal of Contemporary History*, 34:4, 653–62.
Moser, S. and Smiles, S. (2005), 'Introduction: The Image in Question', in S. Moser and S. Smiles (eds) *Archaeology and the Image*, 1–12. (Oxford: Blackwell).
Nora, P. (1989), 'Between History and Memory: Les Lieux de Memoire', *Representations*, 26, 7–25.
Novick, P. (1999), *The Holocaust in American Life* (Boston: Houghton Mifflin Company).
Plato (1973), *Theatetus* (Oxford: Clarendon Press).
Quintilian (2001), *The Orator's Education: Books 11–12* (Cambridge, MA: Harvard University Press).
Renfrew, C. (1994), 'The Identity of Europe in Prehistoric Archaeology', *Journal of European Archaeology*, 2:2, 153–73.
Russell, B. (1921), *The Analysis of Mind* (London: Allen and Unwin).
Russell, I. and Cochrane, A. (2007), 'Visualising Archaeologies: A Manifesto', *Cambridge Archaeological Journal*, 17:1, 3–19.
Samuel, R. (1989), 'Introduction: The Figures of National Myth', in R. Samuel (eds) *Patriotism: The Making and Unmaking of British National Identity, Volume III National Fictions*, xi–xxix. (London and New York: Routledge).
Samuel, R. and Thompson, P. (1989), 'Introduction', in R. Samuel and P. Thompson (eds) *The Myths We Live By*, 1–21. (London and New York: Routledge).
Saunders, N. (2001), 'Matter and Memory in the Landscapes of Conflict: The Western Front 1914–1999', in B. Bender and M. Viner (eds) *Contested Landscapes: Movement, Exile and Place*, 37–54. (Oxford: Berg).
Shanks, M. (1997), 'Photography and Archaeology', in B. Molyneaux (ed.) *The Cultural Life of Images*, 73–107. (London and New York: Routledge).
Sheffield, G. (2002), *Forgotten Victory: The First World War: Myths and Realities* (London: Review).
Smith, A. (1995), *Nations and Nationalism in a Global Era* (Cambridge: Polity Press).
Sontag, S. (1990), *On Photography* (New York: Anchor Books/Doubleday).
Sontag, S. (2003), *Regarding the Pain of Others* (London: Penguin).
Stummer, R. (1998), 'The War we Can't Let Go', *The Guardian Weekend Magazine*, 7 November, 12–23.
Terraine, J. (1980), *The Smoke and the Fire: Myths and Anti-Myths of War 1861–1945* (London: Sidgwick and Jackson).
Todman, D. (2005), *The Great War: Myth and Memory* (London: Hambledon).
Whitehead, C. (2003), *Castles of the Mind: A Study of Medieval Architectural Allegory* (Cardiff: University of Wales Press).
Wilson, R. (2007), 'Archaeology on the Western Front: The Archaeology of Popular Myths', *Public Archaeology*, 6:4, 227–41.
Wittgenstein, L. (1998), *Culture and Value* (Oxford: Blackwell).
Yates, F. (1996), *The Art of Memory* (Harmondsworth: Penguin).

Chapter 6
Historiography and Virtuality

Jerome de Groot

> With modern technology, mere possession of the relics of the past is of little importance. All that *is* of importance is that those entrusted with the care of these fragile and fading things should have the requisite skills – and resources – to prolong their life indefinitely, and to send their representations, fresh, vivid, even, as you have seen, *more* vivid than in the flesh, so to speak, journeying around the world (Byatt 1990, 387, emphasis in original).

In 2004–2005, some four million people visited the Tate website. Over the same period around six million visited the Tate suite of museums physically. Tate Online, the museums' website, is considered the fifth Tate site, with now around a million discrete hits a month. Whilst it is an acknowledged leader in museum online development, the experience of this one organization is instructive and points to a new, expanding and evolving set of heritage paradigms. The internet has meant a clear shift in the ways in which museums are consumed, engaged with and visited. As Ross Parry and Nadia Arbach (2005) have argued, this brings with it an entirely new set of issues for curators and museum managers (as well as for those studying such institutions). The demographic of the audience has changed and their purposes for visiting differ. Importantly, the online visitor is not physically within the institutional museum space and their location (and therefore how they engage with the artefacts, the collection and the experience) is entirely unique to their visit (Stratton 2000, 721–31). Rather than the museum as a *place* being of importance, the way that the collection is *presented* is key. Virtual architecture impacts upon the visitor experience and this has benefits ranging from easier language conversion to faster access for the disabled. The development of virtual museum interfaces is constantly evolving. Key to this development is the shifting definition of the dynamic, context and purpose of the 'visit'. In a very real sense, the online museum has to actively compete with other attractions for the attention of the visitor and must bend to their will in terms of access, route and duration of visit. They are no longer able to physically impart meaning to an object, in the same way that the object is no longer experienced physically in any sense at all. The 'museum' as physical repository is replaced by the cybermuseum, a more negotiable space in all senses of the word. The engagement with the particular artefact is no longer the most important element of the 'visit'.

The impact of this virtuality on the experience of the past is unprecedented. History is always a mediated experience, but the movement away from text and artefact toward virtual or material or physical history presents us with a new and developing historiography. The physical consumption of the historical – economic or otherwise – has been revolutionized. This chapter contributes to a new theorization of the experience of heritage and a new way of understanding the historical by considering the impact of new virtual technologies on the visual and ontological experience of the past (Dicks et al. 2005, Smith 2006). In particular, it considers how performance works within this new set of paradigms. As it demonstrates, the implications of the new visual virtuality to historiography entails a shift in questions of performance and education. If 'the modern museum can be read as knowledge made spatial' and if the 'visit' is the fundamental unit of the user-museum relationship, then this currency has been devalued and changed entirely (Parry 2007, 86).

The chapter demonstrates this by assuming and working through two key ideas: first, the idea that the visitor engages in a kind of 'performance' when going to a museum and that the virtual visitor similarly does this, but in a way that is only really traceable through various softwares; and second, that virtuality provokes a fundamental shift in the concept of 'performance' and the museum, thereby signalling a change in the ways heritage might be performed and interpreted. To demonstrate this, I consider avatars and the use of projected selves in new museum-based games. Both notions of performance involve a virtual/visual interpretation nexus to be overlain on the framework of the visit, re-inscribing this fundamental notion of engagement with the past. In the field of power relations implemented by the newly visual/virtual museum, defined as it is by new, economically inflected web paradigms of representation rather than older tropes, the ways in which the visitor might be expected to perform – or is able to perform dissonantly, for instance – opens up an aperture in heritage which might be suggested to usher in a new way of conceptualizing the past. That said, several commentators on this newly virtual heritage have already suggested that little has changed fundamentally, suggesting a revisionary historiography in which e-tangibles are engaged with, consumed and viewed in much the same ways in which their material counterparts have always been.[1]

Digitization and Visitor Engagement

Webpage and database information exists in a state of simultaneity and potentiality. They are open, unfixed entities that suggest the possibility of navigation away from them. The modern website has a multiplicity of elements including Google Earth tags, blogs and vlogs, images including video and stills, graphics and animation, text and hyperlinks. Museum websites generally eschew such complex rendering, but their information architecture is still as complex and new. The suite of technologies known as Web 2.0, essentially software that allows the user increased control over the content – variously described as hacking or mash-up, and

1 I owe the phrase 'e-tangible' to Ross Parry (2007).

anticipating an e-ontology which articulates itself through collage, disruption and interaction – has led to a great shift in user engagement. In particular, digitization and virtuality have transformed the *experience* of the collection. Mark Whitmore, Director of Collections at the Imperial War Museum, suggests:

> I don't see digitisation as a replacement for the real thing, but rather as another tool for caring for collections and making them accessible. In our case, I think that a visit to a branch is primarily about interacting with the real thing, but that digital material can allow people to potentially delve into areas of personal interest quickly and effectively.[2]

For Whitmore, digitization is a tool that enables faster and more personal access than the public museum space might allow for. This kind of enfranchisement allows the museum to diffract into any number of specific, local experiences. Yet it can also replace an actual visit. Whitmore accentuates the importance of being able to reach people around the world: 'a key aspect of digitization is the ability for us to provide access to remote users (especially in a case like ours where our remit covers the Commonwealth in particular, making visits impractical for many potential users)'.[3] So the virtual visit opens access to global visitations, and in some ways this can be conceptualized as an actual visit. Furthermore, the use of digitized archives can enhance the experience on the ground: 'digitization of related materials can also be used to heighten the appreciation of material in a museum environment (especially functional objects which often need help in communicating their significance)'.[4]

The mode of navigating a museum website is much different to that of a physical space, and so the narrative of the visit is much more dynamic. Furthermore, umbrella sites such as the UK government's 24hourmuseum.org, and search engines like Google Image, emphasize the fact that a visitor will no longer be satisfied with accessing one collection. Rather than looking linearly, users search horizontally; why track through the National Gallery's paintings when a user can look collectively for all works by Holbein on the web? Whilst the bigger collections have the draw and name recognition to ensure a corporate identity, the online visitor and tourist will increasingly desire artefact-led experiences rather than engage with the broader narrative of the museum collection. Museums become content-providers, virtual repositories of information and images, in many ways emasculated and unable to influence the flow, transmission and presentation of knowledge in the ways in which they had once been used. Many museums on the web only exist as cyberentities, having no physical space. The visitor's engagement with, for instance, the Museum of Ephemeral Cultural Artefacts (MECA), is, whilst served from Connecticut, purely online – indeed, the Museum's ethos as

2 Email correspondence with author, 14/11/2005, 16:35.
3 Email correspondence with author, 14/11/2005, 16:35.
4 Email correspondence with author, 14/11/2005, 16:35.

'A cybergallery of transient art and artefacts' suggests that the web manifestation is more accessible (and long-lasting) than the actual objects themselves (whose ephemerality is their point).[5]

Museums can be accessed on normal PCs, through handheld and mobile devices, on phones and laptops. The location of the visitor experience, therefore, is multiple: 'On-line experiences, after all, do not take place in some removed metaphysical *virtual* world. Rather, they are embedded in the actuality and physicality of the user's immediate sense of place, in the same way as the telephone (land line or mobile), newspapers, the radio or TV' (Parry and Arbach 2005). The ontology of the heritage experience shifts. The ways in which the online user performs their role as visitor/tourist is therefore different. Indeed, the entire 'performance' of the museum as physical repository of knowledge and education has shifted. Museums have long been theorized as producers of knowledge (Hooper-Greenhill 1992; Bennett 1995). Part of the processes of this production is the performance of authority and the visitor's role in that dynamic (Hooper-Greenhill 1994). The interface between visitor, institution and artefact online, however, shifts and the museum's ability to work in the standard, traditional ways are lessened; the production of power, or at least the management of power structures and hegemonies, is at the very least challenged or defrayed. The virtual museum is performed, or at least draws attention to its performance of knowledge and power – due to the interpretative and virtual distance between 'real thing' and visitor – in a way that the actual physical museum does not allow for. The museum is always a performance of education and power. The performance of the institution links with certain cultural discourses of education. The online museum gestures to these but lacks the ability to articulate – or to disavow its innate false and incomplete nature.

A key approach to the idea of the 'visit', then, is to think about how users of computers might conceptualize their own relationship to the technology and the internet. How 'embodied' is this interface? The online visit is both purer – insofar as it envisages a direct relationship between visitor and artefact – and noisier – insofar as there are more layers in between viewer and item. This is not particularly unusual – catalogues have replicated images and artefacts for centuries – but the shift is that the representation of the virtual object purports to be an authentic experience rather than an artificial rendering. The online visiting experience suggests an authenticity, a reality or truthfulness in the encounter with the digital artefact that the reproduced image does not. Visually, the online exhibit allows a good engagement with the item. There are no distractions, for instance, and the viewer can gaze as long as they wish at a piece of art, or focus on a particular detail of a sculpture. Images allow for a much closer and more detailed examination of an artefact. The Virtual Egyptian Museum, hosted by the California Institute of World Archaeology has 66 images alone for one of its major exhibits, from large overviews to minutely detailed close-ups. This sense of the artefact being broken into pieces, engaged with both on a wide level and

5 Available at: http://www.edgechaos.com/MECA/MECA.html [accessed: 6 March 2009].

as a fragmented entity, again reconceptualises the relationship of the viewer to the piece.[6] However, it is a digital reproducing of the 'real' art, a virtual simulation which effaces the unique physicality of the heritage artefact. Digitization foregrounds issues of simulacra and artistic disjunction between 'real' and 'rendering'. Is all of this 'virtual' self-presentation 'performance' of a kind? Virtual engagement is constantly a performance, a representation of the physical self/thing. The interaction between the actual museum space and the virtual museum space is a performance, a necessary gesture by one to *become* the other.

Virtuality and Visuality: The Tate

Tate Online, the website of the Tate Galleries, is a massive suite of pages dedicated to creating an online environment for web visitors allows the browser to visit exhibitions online and move around the rooms of the installation. The movement is controlled by interactive mouse, which can be left on autopath, or navigated by the use of hotspots. The physicality of the visit is rendered two-dimensional, and the performance of the visitor is less mediated – controlled (it cannot be random) but possibly disengaged (it can be going on without a viewer). This online exhibiting is relatively common with museums, but for an art gallery it presents a hierarchy of tactility. Work with relief and depth such as thick painting is made into two dimensions, whereas video and photographs are rendered more accurately.[7] Net Art, created specifically for online exhibition, also has the ability to work more effectively. Similarly, performance art lends itself to vlogs and recording, despite the fact that the actuality of embodiment is part of the point of such work. Indeed, video of performances such as John Cage's *Musicircus* on the Tate website are innovative and dynamic, but make the piece – which is interested in simultaneity and immediacy – into something archived and static; video renders it inert and neutral.[8] The experience of this work, then, is deferred through video. Similarly, though, the Tate can also do truly innovative projects relating to visual/virtuality, including their iMap project for blind and partially sighted users, which 'incorporates text, audio, image enhancement and deconstruction, animation and raised images', introducing 'detail in a carefully planned sequence, gradually building towards an understanding of the work as a whole'.[9] This notion of understanding through the piecing together of fragmentary elements is instructive as a pedagogic visual mode, but it also demonstrates the

---

6 Available at: http://www.virtual-egyptian-museum.org/Collection/FullVisit/Collection.FullVisit-FR.html [accessed: 9 May 2008].

7 Available at: See Peter Doig's work, http://www.tate.org.uk/britain/exhibitions/peterdoig/video.shtm [accessed: 19 March 2008].

8 Available at: http://www.tate.org.uk/modern/thelongweekend/cage/video1.htm [accessed: 19 March 2008].

9 Available at: http://www.tate.org.uk/imap/ [accessed: 9 May 2008].

potentiality of the online environment to fundamentally re-inscribe and reinterpret works for particular audiences.

For the Tate, as a gallery of modern art that increasingly includes performance, site-specific installation and video, being online means an archiving of the works (also a restaging of works to then be recorded for the archive). The Museum has instituted the *Tate Player*, a suite of interactive video and audio interfaces including interviews, vodcasts, archived work, and performance (particularly video/live).[10] There are also brief videos entitled *Tateshots*, which the user can subscribe to (RSS) as MP4 files downloaded directly to iTunes or other media playing software (and also download transcript as text). Part of the archive is a set of videos of artists describing their works and methodology, and incorporating within them various works of performance art. The videos also include artists introducing other artists and their own work, even whilst installing it. This blurring of the interface between viewer and artist is unusual in museum settings. The *Tateshot* for the conceptual artist David Lamelas introduces a work that is relatively new in its manifestation: 'Though the project was created in 1970, it was only recently acquired for the Tate Collection, one of the first examples of Tate buying a performance in the form of a set of instructions explaining how to restage it, rather than a physical object'.[11] He argues that he is 'able to produce a work that can exist as a concept rather than an object' – 'Time', the piece being performed, 'is about the mortality of art'.[12] This performance comments upon its own ephemerality whilst it is being videoed and therefore archived.

The Tate's virtual manifestation bears two key imprints of economic compromise. One is the use of technical tools – the Tate's site is driven and built by British Telecom, and each page has their corporate logo and the motto 'Bringing Innovation & Technology Together'. The online museum is produced – literally – through a partnership of corporate technical muscle and artistic content. The Tate's partnership with BT has resulted in multiple awards and has allowed for a development that the museum might not have been able to afford otherwise. The organizations claim that their 'online partnership helps widen access to the arts and engage new audiences', and this rhetoric of access and development again points to an evolving set of paradigms for assessing the effectiveness of a museum in its online manifestation, for adjudging the interface between museum institutions and business organizations, and for conceptualizing the ways in which these 'new audiences' engage with art, heritage and culture.[13] Of course, 'new audiences' is an idealistically vague formulation and ignores many of the worrying hierarchical issues associated with the new information society, insofar as access to the internet is still unequal and subject to modalities of class, economics and culture.

---

10 Available at: http://tate.org.uk/tateplayer/ [accessed: 26 March 2008].

11 Available at: http://tate.org.uk/tateshots/episode.jsp?item=14528 [accessed: 19 March 2008].

12 Available at: http://tate.org.uk/tateshots/episode.jsp?item=14528 [accessed: 19 March 2008].

13 Available at: http://www.tate.org.uk/supportus/corporate/bt/ [accessed: 9 May 2008].

The most obvious economic element of the website, though, is the higher profile of the shop. Rather than being a physical part of the museum complex that could be avoided, the online shop is an inherent part of the information architecture of the website. The online shop allows the visitor to engage in the most important of museum-related branding performances: the purchasing of product and artefact-commodities. The shopsite undermines the visual and virtual hierarchies of the rest of the online museum, as it presents manifestations and simulations of artefacts. The branding of objects with art or the Tate's logos apportions them extra-intrinsic value and the owner cultural capital. Through the purchase of the aesthetic – although an art which is mass produced and commodified – the online visitor strives to attain economic completeness. This is then carried – often literally – into the 'real' world as the visitor's cultural capital may be deployed and demonstrated through their use of various museum-commodities (similarly the National Archives allows the user to send images as e-cards, deploying the artefacts as their own virtual possessions). If we conceptualize the visit as a performance, then the engagement with objects such as this renders the consumption of heritage as something new and extremely interesting. If the online museum re-scripts the visit, the online shop might tend to recast the visit in new terms; indeed, to define the visit purely economically, and to allow it to entirely circumvent the point of the museum (its physical collection) in favour of the purchasing of commodities ascribing the buyer cultural capital without the necessity for anything so awkward as an intellectual or aesthetic engagement.

Virtuality and Community

One of the key elements of the new online museum is its deployment of new marketing, organizational and webgeographic tools to locate itself in the public consciousness. The landscape of the web demands that the museum situates itself in multiple ways using innovative and complex software and technologies. The authority of the museum as institution begins to fray in these new contexts. Facebook, the social networking site, includes various tools and applications to enable members of the site to view and share museum content. The application Museum Video Podcast Player (MVP), for instance, automatically updates the user's homesite with new videos from museum sites globally (including the Guggenheim, Tate, Getty). Subscribers will therefore have museum blogs, content and videos as part of the decoration on their Facebook homepage, and will share them with their friends as part of their virtual social performance. The National Archives similarly allows users to virtually bookmark pages using folksonomy/ social networking software (Facebook, Bebo), as well as information software (Digg) and blogware (Reddit). These social networking functions allow users to share the page/information/artefact with contacts, adding a new level of interaction and community to the dynamic. This social bookmarking, together with more interactive information tagging, emphasizes that the museum is now part of a broader folksonomy, and therefore its unique claims to legitimacy and authority are being eroded. The term 'folksonomy', in fact, might be placed in deliberate tension with the word 'taxonomy', being a way of arranging and collecting information that is anti-hierarchical and horizontal.

At a more fundamental level, museums use networking interfaces to market and present themselves. The Solomon R. Guggenheim Museum in New York, for instance, has a Facebook presence with c.18,000 members. The Facebook site has links to exhibitions, fliers, podcasts, RSS news feeds and publicity. The experience, and performance, of visiting in this instance is equated with being a 'fan' – Facebook's term for someone who is associated with an institutional space – and thence becoming globally networked through the museum. The institution sends updates to 'fans' via the Facebook site, and, crucially, a revolving number of fans – and therefore their image-links – are on the Guggenheim Facebook site, a link to which appears on the fan's homepage. This has implications for a sense of community – the visitor experience becomes something more to do with association, branding and public self-representation (Rheingold 2006; Affleck and Kvan 2008). The fan gains social and cultural capital from being associated with the museum; the institution becomes something much more vaguely defined, particularly in its purpose. It is part of a nexus of community, sociality and database networking, an economic space in which it is marketing itself. The museum is also creating an image, or representation of itself online in this forum – Facebook is not controlled by the institution, it is something that the museum complies with and becomes a member of. The museum performs social networking, presents itself within the forum of virtual interaction and web interface. Its status as an institution is compromised. The National Archives Facebook site, for instance, has a discussion board and public 'Wall' on which users and members post messages. The museum here is part of a wider internet phenomenon of discursivity and interactivity, abstracted from its physical entity and its material collections. This virtual-physical manifestation of the institution fundamentally re-inscribes its relationship to wider culture and to the specific users and fans that form part of this network.

Another area where such social virtuality has enfranchised the visitor is the deployment of user-generated content. Museum-related blogs, vlogs and YouTube videos are commonplace, with entire communities coming into being to discuss, argue and debate museums. Amateur users make videos about their experiences, performing a visit. Likewise, professional filmmakers and curators make films and circulate them using such sites in order to facilitate learning and publicize particular exhibitions. Any visit can now be uploaded, as videos taken whilst touring the museum are put online. These videos are then shared, added to blogs and embedded into personal webpages. Amateur video on YouTube and video sharing sites can then create a forum for visitor-led discussion, including – for instance – an exchange around a video uploaded of Cairo museum on YouTube.[14] The video provoked a

14 Available at: http://youtube.com/watch?v=WJ0p7NPoTrM [accessed: Feb 10 2010].

range of reactions, including gratitude ('Thankyou for that video. I've never been so that was nice to see, but looked hot and crowded') and hostility ('a few selfish, thoughtless morons don't understand the meaning of No Cameras').[15] The first suggests that virtual tourists might use such amateur films to engage with images of museum experiences while remaining aware of their physical distance (and possibly be happier not to be in the 'hot and crowded' museum). The second raises a curatorial issue about access and conservation. Both comments demonstrate the new ways in which user-generated information might be interpreted and used by the global online community. They are also suggestive of a shift in the performance and actions of the visitor, from interested observer to educator, cultural mediator and enabler. The embodied experience of the filmmaker is referred to by another post: 'lol wtf is with that guy or gal shining the light right into your eyes when you were looking in that one case?'[16] These contributions reveal the new ways in which the 'visit' itself is being conceptualized, broadcast and critiqued within online communities. In order to maintain their status within such confluences of discourse they do not control, institutions have to work very hard, as the visual or representational tropes now associated with the museum are fracturing.

Avatars and Virtual Reality

This fracturing is demonstrated most clearly by the advent of museums in virtual worlds, and the concomitant bleeding back of gaming models into museum online representation. The deployment of an avatar in a virtual representation defers the individual's engagement with 'reality', at once opening up the possibility of multiple global relationships and dissident performances of self, whilst simultaneously abstracting the player from the material world. The use of Virtual Reality (VR) landscapes in museum websites demonstrates the extent to which the defrayal of the museum 'visit' which I have been describing so far has come. The 'performance' of the visitor is now uppermost in the experiential nexus, but in simultaneity distanced from any kind of mainstream definition of normality.

Second Life is the biggest virtual world in which avatars interact with a created landscape and each other.[17] Various virtual museums have been built, including the Second Louvre, the Open Source Museum of Open Source Art, and the Star Trek Museum. The various museums have memberships, ranging from a handful to 889 for a Dutch Masters museum which replicates in 3D the paintings of Van Gogh (di Blas et al. 2005). This museum allows the user to take an image from an online database and recreate it in three dimensions to the extent of 'bodily' entering it via

15 Magaboosa, posted 18 March 2008, farfar5010, posted 22 February 2008, available at: http://youtube.com/watch?v=WJ0p7NPoTrM [accessed: 25 March 2008].

16 Slvgdvg, posted 21 February 2008, available at: http://youtube.com/watch?v=WJ0p7NPoTrM [accessed: 25 March 2008].

17 Schultze and Rennecker (2007) provides a useful overview.

their avatar. The Second Life museums have events including community visit days, ranging from the educational to the fantastical:

> The Star Trek Museum invites you to visit and swim with George and Gracie. These humpback whales were brought here from the 20th century by Admiral James T. Kirk in an effort to re-populate the species. They can be viewed from the observation area to the south of the Star Trek Museum. Museum Staff will be available to answer your questions.[18]

This participatory-fantasy virtual experience presents the museum as somewhere to interact, engage and virtual-bodily interface with (Urban et al. 2007). The sense of the museum as something *participatory* is key. The avatar-projection is able to interface and interact with the object, to perform interrelationship with it – to even, if they wish, recreate and perform within it. The visual tropes associated with the museum are fractured and interrogated by a community happily destroying sources of cultural legitimacy.

To counter this site non-specificity, traditional museums have various strategies. Living history sites emphasize their location-specific importance, suggesting that in contrast with the anonymised virtual access that a user might undertake in a visit to a traditional museum, it is still important to bodily *experience* the past. However, virtual reality technology is increasingly used to replicate the past for the user and it is clear that this type of experience will be of great use to, and influence on, living history approaches. Museum theatre again seemingly emphasizes the 'now-ness' and physical moment of the visitor's engagement with past and performance. The use of vlogs, video and livefeeds might allow theatre to become part of the online museum experience, but it will no longer be 'theatre' as traditionalists might conceptualize it. However, dramatic adjunct and interpretation is more suited to the modern online experience – the YouTube generation expects performance interpretation and some kind of dynamic visual interface which the ordinary, inert artefact cannot render. Indeed, performance becomes more embedded in the online visitor experience, something much more associated and integrated with the rest of the site and therefore its ability to make other and strange undermined but its legitimacy heightened. Museums, then, might respond to the new visiting strategies of the network society by emphasizing the important performances at the heart of their project.

Institutions have responded to the technical and physical challenge of VR worlds by developing their own gaming interfaces. *Discover Babylon* is a joint project between various organizations including the University of California, Los Angeles (UCLA) Cuneiform Digital Library Initiative, the Walters Art Museum, the Federation of American Scientists and Escape Hatch Entertainment (Lucey-Roper 2006). Such synergy and combination of particular skills – curatorial,

18 Available at: http://secondlife.com/events/event.php?id=1302881&date=1205452800 [accessed: 25 March 2008].

educational, technological, economic – is common in virtual projects due to the range of technical demands, and as such, the development of virtual technologies in museums emphasizes a new interdisciplinary approach to curatorship and representation. Discover Babylon aims to use 'sophisticated video gaming strategies and realistic digital environments to engage the learner in challenges and mysteries that can only be solved through developing an understanding of Mesopotamian society, business practices, and trade'.[19] The project conceptualizes the past as an immersive game, something to be engaged with and consumed within a particular set of leisure time contexts – education by stealth in many ways. The gameplay mimics Second Life in its visuality, consciously echoing another mode of virtual interaction (Rejack 2007).

The artefacts in the gameplay are those found in the Walters Art Museum in Baltimore. However, rather than simply allowing the avatar to tour the museum, the game deploys a cataclysmic doomsday scenario familiar from such games as *Resident Evil* and particularly *Tomb Raider*. A maverick archaeologist has damaged the fabric of time and the player has to undertake a series of missions in order to fix this. The invocation of *Tomb Raider* demonstrates the way that the avatar-performance in the game takes on a number of layers. *Tomb Raider* – the game – and its various film manifestations introduced the world to Lara Croft, adventurer-archaeologist (de Groot 2008, 49–53). The deployment of a set of gaming tropes attempts to make this part of a globally embodied historical experience, in the manner of such Massive Multiplayer Online Role Playing Games as *Rome: Total Victor*. *Discover Babylon* enables the player to inhabit a variety of historically attested characters, performing their historicity virtually (whilst moving them in an ahistorical electronic environment). The avatar-projection of selfhood into history is compelling here – whilst the VR element of the game obviously distances the player from the past, they are expected to actively engage pseudo-bodily with the artefacts and historical environment (de Groot 2006). The game therefore raises various questions of historical empathy, experience, and embodiment, as well as self-projection. It is ergodic and based around a set of tasks, casting the past as a narrative in which education leads to understanding and therefore progression. It suggests that skills associated with video gameplaying – problem solving, virtual conceptualization, interpretative analysis, plan formulation, adaptation – are all worthy of encouraging and supporting through the museum interface.[20] Indeed, it suggests that this ludic and embodied approach to presenting the heritage collection is fundamentally interactive and educational.

---

19 'Discover Babylon', available at: http://www.discoverbabylon.org/index.asp [accessed: 25 March 2008].

20 See the FAS 2005 report 'Summit on Educational Games' which recommended that research into educational games be part of the US government's education strategy, available at: http://www.fas.org/gamesummit/Resources/Summit%20on%20Educational%20Games.pdf [accessed: 26 March 2008].

Similar projects are being finalized including relationships between Shakespeare's Globe and Second Life performers, and the *Middleton Mystery* game developed by Belsay Hall, Northumberland, in conjunction with a set of school students. These increasingly hybrid developments demonstrate the multiple ways in which museums are developing their online profile. The museum is still part of a discourse of visual spectacle, of a nexus of power relations between viewer and artefact, and the repository of national narratives and institutional power. Yet all of these relationships and phenomena are troubled, interrogated, reanimated and complicated by the advent of increasingly complex technologies and, concomitantly, the development of a visitor body increasingly confident in a multiplicity of heritage experiences and engagements. Since their inception as institutions in the late eighteenth century, and even from their origins in collecting and display during the early modern period, museums have always being imagined spaces for the purposes of virtual visiting and experience. They allow the user to engage with culture, history and geographies that are other to their daily experience, without leaving the relatively comfortable confines of their house, home city, or country. Visitors to museums experience artefacts that are othered by distance, space and chronology; to tour the culture and history of the world. The museum has always, then, been a space for the imaginary reality, for a fantastical self-projection and an interface with whatever the artefact is metonymic for – history, ethnography, culture, society – that is predicated upon that object performing a visual narrative to the visitor; what it *means*, then, as a piece, enables the visitor to experience vicariously/virtually. New media practices and technologies allow this function of the museum to continue, but – crucially – without the museum as institution being as authoritative in its presentation of the various narratives as previously (Clifford 1997). If the space of the museum is no longer able to impose meaning, to frame the visit and to impress (bodily and conceptually) the visitor with information and focus, then how will heritage institutions maintain their hold over the national imagination? The various technologies I have outlined here seemingly enfranchise the visitor to make their own narrative from the building blocks of a museum's collection, to intervene with individual stories and devolve power away from the central institution. If museums can no longer control their visitors, and if those members of the public decide to perform 'visit' in new and dynamic ways, an entirely new set of relationships begins to evolve.

References

Affleck, J. and Kvan, T. (2008), 'A Virtual Community as the Context for Discursive Interpretation: A Role in Cultural Heritage Management', *International Journal of Heritage Studies* 14:3, 268–80.

Bennett, T. (1995), *The Birth of the Museum.* (London: Routledge).

di Blas, N., Gobbo, E. and Paolini, P. (2005), '3D Worlds and Cultural Heritage: Realism vs. Virtual Presence', in J. Trant and D. Bearman (eds) *Museums and the Web 2005: Proceedings* [Online: Toronto: Archives & Museum Informatics]. Available at http://www.archimuse.com/mw2005/papers/diBlas/diBlas.html [accessed: 25 March 2008].

Byatt, A.S. (1990), *Possession.* (London: Vintage).

Clifford, J. (1997), *Routes: Travel and Translation in the Late Twentieth Century.* (Cambridge, MA: Harvard University Press).

de Groot, J. (2006), 'Empathy and Enfranchisement: Popular Histories', *Rethinking History: The Journal of Theory and Practice* 10:3, 391–413.

de Groot, J. (2008), *Consuming History.* (London: Routledge).

Dicks, B., Mason, B., Coffey, A. and Atkinson, P. (2005), *Qualitative Research and Hypermedia: Ethnography for the Digital Age.* (London: Sage).

Hooper-Greenhill, E. (1992), *Museums and the Shaping of Knowledge.* (London and New York: Routledge).

Hooper-Greenhill, E. (1994), *Museums and their Visitors.* (London and New York: Routledge).

Lucey-Roper, M. (2006), 'Discover Babylon: Creating a Vivid User Experience by Exploiting Features of Video Games and Uniting Museum and Library Collections', in J. Trant and D. Bearman (eds) *Museums and the Web 2006: Proceedings* [Online: Toronto: Archives & Museum Informatics]. Available at http://www.archimuse.com/mw2006/papers/lucey-roper/lucey-roper.html [accessed: 26 March 2008].

Parry, R. (2007), *Recoding the Museum.* (London and New York: Routledge).

Parry, R. and Arbach, N. (2005), 'The Localized Learner: Acknowledging Distance and Situatedness in On-Line Museum Learning', in J. Trant and D. Bearman, (eds) *Museums and the Web 2005: Proceedings* [Online: Toronto: Archives & Museum Informatics]. Available at http://www.archimuse.com/mw2005/papers/parry/parry.html [accessed: 1 June 2006].

Rejack, B. (2007), 'Toward a Virtual Re-enactment of History: Video Games and the Recreation of the Past', *Rethinking History* 11:3, 411–25.

Rheingold, H. (2000), *The Virtual Community.* (Boston, MA: MIT Press).

Schultze, U. and Rennecker, J. (2007), *Refraining Online Games.* (Boston: Sprinter).

Smith, L. (2006), *Uses of Heritage.* (London and New York: Routledge).

Stratton, J. (2000), 'Cyberspace and the Globalization of Culture' in D. Bell and B.M. Kennedy (eds), *The Cybercultures Reader.* (London and New York: Routledge).

Urban, R., Marty, P. and Twidale, M. (2007), 'A Second Life for Your Museum: 3D Multi-User Virtual Environments and Museums', in J. Trant and D. Bearman (eds) *Museums and the Web 2007: Proceedings* [Online: Toronto: Archives & Museum Informatics]. Available at http://www.archimuse.com/mw2007/papers/urban/urban.html [accessed: 25 March 2008].

Chapter 7
Visualizing the Past: Baudrillard, Intensities of the Hyper-real and the Erosion of Historicity

Richard Voase

This chapter aims to bring specificity to the links between the hyper-real and visualizations of the past. Theoretically, it draws on the work of Jean Baudrillard, particularly his exploration of the consequences of the proliferation of transmitted information. Here, he argues, technological change has created a world sated with informational messages, in which everyday life is increasingly lived through mediated images. These images sometimes faithfully represent the real; but commonly, to attract attention, engage and entertain, they simulate the real in varying levels of intensity. The result is beyond real: it is a hyper-real. While this term has had a tendency to be deployed rather loosely, in this chapter I show that Baudrillard's writings, admittedly opaque, offer a model by which the hyper-real can be understood in specific contexts. For example, by applying the model to particular cases in the fields of cinematic film and heritage interpretation, different intensities of the hyper-real can be diagnosed; and as shall be seen, historicity is compromised.

Before proceeding further, it may be useful to explain the term 'historicity' as used in the context of this chapter. Historicity, here, is intended to mean genuineness, in the sense specified in the Oxford Dictionary. There is, of course, a debate about the 'genuineness' of history. History is itself an interpretation of the past, based on an assemblage of surviving sources (Lowenthal 1998, 112–3). Items such as newspapers from a period are treated by historians as primary sources (Marwick 1989, 199); but the writings in those newspapers are, themselves, interpretations of the times, albeit written by contemporary interpreters; and interpretation is inseparable from value judgements (Carr 2001, 101). History, itself, is an *histoire.* To debate this further would distract from the intended focus of the chapter, but suffice it to say that the member of the public who visits a museum, or who watches a film, may be presented with claims to historicity – genuineness – of varying intensity. On some occasions, those claims may extend to explaining the present as a result of causal processes emanating from the past (Marwick 1989, 398). This chapter proposes a method for diagnosing the extent of deviance, whether unintended or deliberate, from that which can be reasonably regarded as 'genuine'.

It should be stated at the outset that the hyper-real is not solely a product of recent technological advances. Arguably, the story begins in the early decades of the twentieth century, with the proliferation of the cinema and the mass-circulation of magazines bearing good-quality photographs. Lyotard observes how these developments initiated a mass-transformation of the real: '... photographic and cinematic processes can accomplish better, faster, and with a circulation a hundred thousand times larger than narrative or pictorial realism ... This is the way the effects of reality, or if one prefers, the *fantasies of realism*, multiply' (Lyotard 1984, 74, author's emphasis). Thus, the cinema, and photography, take the real and create fantasies from it. Aptly, therefore, the search for illumination begins with the consideration of a particular work of cinema, whose very subject matter concerns the reality of the real.

The Matrix

'The real is produced from miniaturized cells, matrices, and memory banks, models of control – and it can be reproduced an indefinite number of times from these' (Baudrillard 1994, 2). Baudrillard's *Simulacra and Simulation* was first published, in its original French, in 1981. His point, put simply, was that technological advances were amplifying the capacity to circulate information, and this proliferation of information was re-shaping the nature of what one might call 'the perceived real'. This understanding of technology, information proliferation and 'the perceived real' underpins, for example, the science-fiction film *The Matrix* (1999), which owes inspiration to Baudrillard's 1981 publication (Muri 2003, 88). The plot is based on the premise that the perceived real is a fictional artifice, by which the 'real' real is concealed. As the plot develops, perceived reality is revealed to be an illusion, inserted into human minds by machines which, having been created by humans, subjugated them, and used their body heat as a source of bioelectricity (see Lee 2005 for a plot discussion). The machines maintain their human captives in fluid-filled pods, in a physical state of unknowing, immobile passivity. That which the humans understood and experienced as the real world was, in fact, a set of computer-generated illusions: 'The Matrix'. An esoteric visual clue is used by the writer/directors of the film to acknowledge their inspiration from Baudrillard. We first meet the film's protagonist, computer-hacker Neo, in his room. Expecting a visit, he answers the door and then moves back into his room to retrieve a disc of data. Neo opens a book; but its appearance masks its real identity, as it is, in fact, a box, a repository of illicit information held on discs. The text on the cover of the book/box, fleetingly visible, is *Simulacra and Simulation*, a clue that is easily understood by the initiated (Constable 2006, 233).

The Matrix has the appearance of a work of fiction based on the notion that the real is not real. However, in accepting this we become victims of a double bluff. Consider an objective look at what happens in an encounter with this film: we are invited to enter a cinema, a darkened room from which all other visual stimuli are excluded. We are maintained in a state of immobile passivity for the duration of the film. During that time, we are fed with electronically-mediated images and sounds. It is true to say that, for that period of time, we live out our lives in a fictional world. *The Matrix* would have us believe that 'The Matrix' is the result of a nefarious plot by renegade robots that turn to bite the hand that feeds them. That is a common enough storyline in science-fiction. But Baudrillard's 'matrix', in *Simulacra and Simulation*, is *real. The Matrix* – the film – presents itself as fiction, concealing Baudrillard's point that what he refers to as 'miniaturized cells, matrices and memory banks', and their effects, are real. Humans created them and are subjugated by them.

Curiously, it is the initiated who understand the references to Baudrillard and who are potentially the biggest victims of this double bluff. Perhaps self-satisfied by our understanding of the esoteric clues to the source of the film's inspiration, we are diverted from seeing that the book and the film are making diametrically opposed points. *The Matrix* works as The Matrix should: 'It is ... the map that precedes the territory ... that engenders the territory' (Baudrillard 1994, 1). It is an example of what Constable refers to as Baudrillard's 'reversible logic of the hyper-real' (2006, 240). Indeed, what this example illustrates, I hope, is that there is more to the hyper-real than meets the eye. What is needed, therefore, is a method for diagnosing the hyper-real in terms of its subtleties and intensities. Conveniently, Baudrillard provides the tools.

Baudrillard and the Four Phases of the Image

Baudrillard's operating definition of hyper-realism suggests that it is '... the meticulous reduplication of the real, preferably through another reproductive medium ... Through reproduction from one medium into another the real becomes volatile ...' (Baudrillard 1993, 71). In other words, the hyper-real occurs when a 'real' is reproduced as a convincing sign system; that is, a simulation. This may be a faithful reproduction, such as straightforward photography. However, photographs can be retouched or digitally altered. Similar liberties can be taken with other reproductive media. This is the point at which the situation becomes 'volatile'. Things become especially complicated when one simulation is reproduced as another simulation. For example, the British television soap opera, *Coronation Street*, is a drama based on life in a fictional street in Manchester. It is a simulation. It is not possible to visit the 'real' Coronation Street. Or rather, it is. At a Granada Studios Tour, paying visitors can walk around the 'real' Coronation Street. Likewise, visitors to a Disney theme park are invited to visit the 'real' Sleeping Beauty's castle. Both provide examples of 'real' places that are based on fictional originals.

Baudrillard offers specific principles regarding the workings of the hyper-real (1994, 6). He terms these the 'four phases of the image', with each incremental phase labelled level one through to four. The following explanations derive from the version of the text published in the anthology of Baudrillard's writings edited by Poster (2001, 173). The level one phase of the image is said to be 'the reflection of a basic reality'. This can be considered as pure representation. It is an iconic sign, which, though not equivalent to the reality that it represents, faithfully reproduces it. For example, a person's passport photograph is an iconic sign, which faithfully represents his/her appearance. The level two phase of the image, Baudrillard suggests, 'masks and perverts a basic reality'. The author recalls a television advertisement where two women, drinking coffee in a kitchen, hear a knock at the back door. It is a known tennis celebrity who joins them for coffee. The setting is ordinary enough, but the event is improbable. The level three phase of the image 'masks the absence of a basic reality'. This can be illustrated by reference to the website of the Omega Watch Company (Omega 2008). On the site is a list of product 'ambassadors', that is, celebrities contracted for product endorsement purposes. Readers are invited to spot the odd one out:

Cindy Crawford;
Nicole Kidman;
Ellen MacArthur;
Michelle Wie;
Abhishek Bachchan;
Dean Baker;
James Bond;
Eugene Cernan;
George Clooney;
Michael Phelps;
Alexander Popov;
Michael Schumacher;
Ian Thorpe.

These are internationally known screen actors and personalities from the world of sport, with one exception: James Bond is a fictional character. The inclusion of his name in a list of 'real' people serves to mask the absence of a basic reality: he does not actually exist. It can be understood why the company should choose, in this case, the character rather than the actor for the ambassadorial role, but it is a subtle example of the workings of the hyper-real. Baudrillard (1994, 2) observes that '... the era of simulation is inaugurated by a liquidation of all referentials – worse: with their artificial resurrection in the systems of signs ...' In other words, referentials – the realities to which names and images are attached – are less important than the attention that the sign can command. In the case of George Clooney, it is the actor's name and symbolic associations that command attention. In the case of a character such as James Bond, who has been played by several male leads, the symbolic associations are vested in the character rather than the actor.

The final level, Baudrillard's level four phase of the image, in his words, 'bears no relation to any reality whatever; it is its own pure simulacrum'. The level four phase is perhaps the trickiest to translate into an operational understanding, because every image created by the human mind is *de facto* derived from images stored in that mind. For example, science fiction as a cinematic genre should, theoretically, depict images of things and places yet to come, or yet to be experienced. Yet all the fantastical creations in the *Star Wars* series of cinematic films were based on images from the built and natural world on Earth. Baudrillard's intended meaning of the term 'pure simulation' is that it is the consummate simulacrum, wherein, to adapt Baudrillard's own metaphor, the map (as sign) is so convincing it 'engenders the territory' it is supposed to represent, rather than vice versa (Baudrillard 1994, 1). An anecdote may serve to illustrate here. Some years ago, the novel *Possession* by A.S. Byatt was adapted for the cinema, with the actress Gwyneth Paltrow cast in one of the leading roles (*Possession* 2001). Part of the filming took place at the University of Lincoln, where the author works. When the film was released, the university and a local cinema organized a Premiere, complete with a Gwyneth Paltrow look-alike arriving in a stretch limousine. The author was not present, but was shown the photographs. Discussion with a university officer involved in the organization of the event led to the agreement that the fake Gwyneth seemed, somehow, an improvement on the original. The real Gwyneth, seen around the campus filming, was, like many cinematic stars, quite a slight figure. People onscreen tend to look larger than they are in real life. The look-alike conformed to Gwyneth as 'meticulously reproduced through another reproductive medium' (Baudrillard 1993, 71), rather than to the real Gwyneth. Indeed, the fake Gwyneth was more Gwyneth than Gwyneth: a case of the map engendering the territory.

So, to summarize, a level four phase of the image supersedes, or precedes, any original. Baudrillard's own choice of example to illustrate the difference between level four and the first three levels is that of illness: specifically, the difference between illness that is dissembled, and illness that produces psychosomatic symptoms. If a person feigns illness it is usually fairly obvious that they are faking it. But if they produce symptoms, reality is compromised. As Baudrillard (1994, 3) puts it, '... pretending, or dissimulating, leaves the principle of reality intact: the difference is always clear, it is simply masked, whereas simulation threatens the difference between the true and the false, the real and the imaginary.' To return to the earlier example of James Bond, it is likely that many readers spotted the odd one out in the list. Everyone knows, with a moment's reflection, that James Bond is a fictional character, although the likelihood (and intention) is that the list is not read with reflection, and the fiction is not identified. The difference between James Bond and the various male actors who have played him is clear. When the simulated Gwyneth Paltrow conforms more closely to one's expectations of the real Gwyneth than Gwyneth herself, at that point one begins to wonder what was there in the first place. Of course, in terms of acting ability, we should expect that the fake Gwyneth cannot hold a candle to the real Gwyneth. But in a Western culture that is both technologically advanced and intensely visual, there is the ever-present question, in Baudrillard's words, of 'substituting signs of the real for the real itself' (1994, 2).

The contemporary media phenomenon known as 'celebrity culture' has, at its heart, the manufacture of a culture of fame around otherwise unknown and ordinary individuals, usually through the medium of reality television. The reading matter of celebrity culture – a stable of magazines – runs stories about 'celebrities'. Some celebrities are famous for outstanding talent: for example, Gwyneth Paltrow. Alongside these are stories and pictures of those who have become famous through nothing more than media exposure. Because recognition, and hence the power to sell magazines, is all that matters, the provenance of the fame is unimportant. Once again, in Baudrillard's words, it is a case of the 'liquidation of all referentials' (1994, 2). It is time now to examine means by which this same model can be applied to the ways in which the past is represented and understood through the modalities of image production and reproduction.

Historicity Eroded

'When the real is no longer what it used to be, nostalgia assumes its full meaning. There is a proliferation of myths of origin and signs of reality; of second-hand truth, objectivity and authenticity ... there is a panic-stricken production of the real and the referential ...' (Baudrillard, cited in Poster 2001, 174). Earlier, a contextual meaning for 'historicity' was established, relating to the genuineness of the relationship with the past. What Baudrillard suggests above is that once the need to be recognized supersedes the need to be genuine, desire in the form of nostalgia becomes the driving force behind the consumption, and hence the production, of the past. Here is another list: Captain Cook, Mary Queen of Scots, Robin Hood, William Shakespeare, Catherine Cookson, Wolfgang Amadeus Mozart, and Santa Claus. This time, the invitation to the reader is not to identify the odd one out but to identify that which they all have in common. The answer is that their names have all been annexed by tourist destinations for promotional purposes. In that sense, they are equal and the same. And yet their provenances differ greatly. Some are historical figures. Two are fantasy figures with no real historical provenance. And yet, in the case of both Santa Claus and Robin Hood, attempts to explore and identify the character's historical origins have become an industry in itself (see Jones in this volume). Significantly, both are known for their visual identities as well as their names: Santa in his red robe, Robin Hood as the man in tights. Such is the symbolic value of Robin's recognition, that he is claimed by more than one county in England. Paradoxically, alongside the adoption of anything recognizable in order to attract attention – Baudrillard's 'proliferation of myths of origin' – there is a parallel drive, in Baudrillard's words, 'panic stricken' – to legitimize; that is, produce the real behind the sign, hence the quest to ascribe historicity to legendary characters.

This paradox drives, arguably, what has become known as the heritage industry. In Britain, the 1980s was the decade when the country went somewhat heritage mad. The social commentator Robert Hewison, writing in his 1987 polemic *The Heritage Industry*, spoke of the contemporary nostalgia for the past as 'a sickness that has reached fever point' (1987, 10). Two years later, the then president of the Museums Association, addressing that body's annual conference, was reported as condemning 'the current obsession with heritage', and bemoaning a tendency of local governments to divert funds from traditional museum provision into that new breed of attraction, loosely termed 'heritage centre' (Davies 1989, 41). Now is not the time to become embroiled in definitions, but it is important to understand that the relationship between 'heritage' and 'history' is more tenuous than one would think. At the heart of heritage, as with celebrity culture, is recognition. In recent years, 'the past' has featured significantly in television and cinematic schedules; but not the past as totality, but a greatest-hits version of it. The Romans, Henry VIII and Elizabeth I, for example, can be relied upon to draw a crowd, albeit a certain kind of crowd. This specific idea of 'heritage' is what a particular, and dominant, class of people is prepared to consume and pay for, whether through taxation or entry fees. That said, the growth of interest in heritage, particularly in the 1980s, led to an explosion of interest in the past. This was arguably fuelled, as suggested earlier, by the concomitant pursuit of the real, and was represented in an explosion in the number of new museums opening (Hewison 1987, 24).

Perhaps the paradox can be understood by thinking behind the enigmatic statement of Baudrillard, cited earlier: 'When the real is no longer what it used to be, nostalgia assumes its full meaning'. Once cultural boundaries have been eroded and academic proprietorship of the past has been broken, 'the past' becomes a supermarket of styles to be sequestered, and a repository of the real to be raided. Jameson (1984, 58) speaks of '... a whole new culture of the image or the simulacrum (and) a consequent weakening of historicity, both in our relationship to public history.' That which follows is an attempt, through examination of particular cases, to illustrate how this works out in practice.

Cinema and the Erosion of Historicity

> Myth, chased from the real by the violence of history, finds refuge in cinema (Baudrillard 1994, 43).

Baudrillard argues that history has entered cinematic film as a despotic illusion, a 'pacified monotony' which 'celebrates its resurrection in force on the screen' (1994, 43; see also Constable 2006, 242). Alternatively, one could say that the hyper-real pasts presented in contemporary film are somewhat fun. I do not propose to take sides, but to point out that technological and cultural evolution cannot be reversed; and that it is more agreeable, *pace* Baudrillard, to give in to fun than to despair. Perhaps understanding should be the goal, and that is the purpose of this chapter: to comprehend the levels of intensity at which hyper-real liberties have been taken.

Jameson seems less troubled than Baudrillard about what he sees. Just as new categories of visitor attractions – heritage centres – appeared in the 1980s, Jameson argues that new categories are needed for film that in some way harness the past and appeal to nostalgia. *Star Wars*, mentioned earlier, is seen by Jameson as a nostalgia film because its plots echo the sci-fi B-movies of earlier decades, with plucky heroes, damsels to be rescued, and transparently bad characters (1985, 116). One could add, for example, that the design of the 'imperial walkers', the tank-like machines whose forms emulate varieties of dinosaur, also recreate a childlike fascination. This is an emerging category of film, neither historical pageant nor historical epic, which draws on the past in a variety of forms to create pleasing dramas and spectacles.

The Matrix (Revisited)

It is in identifying intensities of the hyper-real in particular films that the four phases of the image become important. Consider again *The Matrix*. The film suggests that the perceived real is an elaborate simulation. It is a fantasy of the 'what if' kind. To borrow and adapt Baudrillard's phraseology (1994, 2), there is a 'sovereign difference' between the film as work-of-art and known reality, and therein lies the 'charm' of fantasy. It appears to fit within phase level one. However, as has been shown earlier, *The Matrix* is a double-bluff. Its inspiration, Baudrillard's *Simulacra and Simulation*, suggests that 'the matrix' is real. *The Matrix* positions us as passive consumers, lapping up apparent fantasy, while its inspiration is a statement on the effect of the proliferation of mediated information on perceived realities. Baudrillard's intended point, that our perceived real has become, in a very real sense, fantasy, is concealed, rendering *The Matrix* a level three simulation.

Sky Captain and the World of Tomorrow

The fantasy adventure *Sky Captain and the World of Tomorrow* (2004), like *Star Wars*, harks back to Buck Rogers' B-movies of the inter-war years. Unlike *Star Wars*, its confection includes *film noir*, artwork redolent of pulp sci-fi magazines of the 1930s, more than a nod toward the James Bond genre, with a mad German scientist played in hologram version by the long-deceased Laurence Olivier, and action sequences in exotic locations around the globe (for a review see French 2004). The film is presented in a tinted monochrome, making, for example, the bright red lips of Gwyneth Paltrow's newspaper reporter, a Lois Lane equivalent, the more noticeable. *Sky Captain*, in Jameson's categorization, is a nostalgia film that takes liberties with the past in order to make itself look good. The opening sequence shows a huge airship making its way through a wintry sky towards the skyline of New York. As it passes, we read the name of the craft in German gothic characters on the side: Hindenburg III. Immediately, informed viewers – and not all are informed – are alerted to the fact that games are being played with history. There was only one Hindenburg, and this famously crashed in New York in 1937. There are no Nazi emblems on the craft; the Third Reich has not happened. This is fantasy, in which the 'reality principle' remains intact. At this point, *Sky Captain* is a level one simulation.

The Hindenburg III comes in to dock, not at an airfield, but atop a skyscraper as docking mast. As it approaches, a door falls open and a rope falls out. A uniformed man, awaiting the airship's arrival, takes the rope and pulls the craft toward him. We have a subsequent glimpse of this character, holding the ship as the passengers disembark. This is a liberty with history which would have produced howls of laughter from a 1930s audience. Though the invention of the docking mast enabled airships to dock without a huge ground crew – a ground crew of 60 men was assembled to handle the Graf Zeppelin for a stopover in Bern in 1930 (Bern Info 2008) – one man could not hold down the Hindenburg, which was almost as large as the Titanic. Nor would the top of a skyscraper serve as a suitable docking mast.[1] The scene provides an impressive introduction to the film, but relies on the audience's detachment from the reality of airships in order to not appear derisory. At this point, *Sky Captain* is a level two simulation.

As the film progresses, the action and the capabilities of the machines belong increasingly and more intensely to the realm of fantasy. Near the end, we are treated to the spectacle of huge aircraft-carrying airships. I thought this a fantastical idea until discovering, while researching this chapter, that the United States had manufactured two aircraft carrying airships: a case of truth being, if not stranger than fiction, rivalling it. Given that the director offers clear visual clues at the beginning of the film that liberties are being taken with history, reality and fantasy are clearly separate. Despite the remarkably strong man holding the Hindenburg down single-handed, fantasy retains its magic. This cannot quite be said about the example that follows.

King Arthur

> Something has disappeared: the sovereign difference, between one and the other, that constituted the charm of abstraction. Because it is difference that constitutes the poetry of the map and the charm of the territory, the magic of the concept and the charm of the real (Baudrillard 1994, 2).

King Arthur (2004) is a reworking of the Arthurian myth. Set around Hadrian's Wall, the former Roman frontier in northern England, Arthur is depicted as a Roman officer. His demobilization is suspended in order for him, and his colleagues, to perform a rescue deep inside Pictish territory, north of the Wall. It is an enjoyable action feature, its plot somewhat redolent of *The Magnificent Seven*. The appearance of elements of Dark Age history such as St Germanus, the Pelagian heresy and the battle of Mons Badonicus, reminded me of an encounter with them during my undergraduate studies (for a good account see Alcock 2001). But in this film, the charm is somewhat squeezed out by a claim, conveyed by text at the beginning of the film, for historical accuracy. First, a (correct) statement claims that there is agreement on a Dark Age provenance for a likely historical origin for

1 I acknowledge the helpful observations of my colleague Peter Lyth, of Nottingham University, regarding the docking of airships in the 1930s.

the Arthurian tale. Viewers are then told that 'Recently discovered archaeological evidence sheds light on his true identity', which is simply not the case.

This antagonized one commentator: 'Strip the tale of King Arthur of its chivalry and sorcery and what is left is nothing short of a national insult', thundered *The Guardian* (Jones 2004). The film's historical consultant, interviewed for the *Edinburgh Evening News*, admits that the film's director had taken liberties with history, but sets out his case for an historical provenance for Arthur as an officer on Hadrian's Wall reflecting several attempts to do this over a long period of time (Stuart-Glennie 1869; Skene 1988 and latterly Moffat 1999). Set against this are comments from an Edinburgh University historian, who finds the evidence slight (Fettes 2004). Claims to locate Arthur in the North, away from the South-West of England where legend and the sketchy available evidence has traditionally located him, does rather smack of Yorkshire disputing with Nottinghamshire over who owns Robin Hood (Yorkshire Robin Hood Society 2002, see Jones this volume).

In hyper-real terms, at what level can *King Arthur* be positioned? Touches such as the use of the hand-held crossbow several centuries before its adoption in Europe would locate it at level two: a perversion of the real. If the account of Alcock (2001) and other academic authorities are accepted, that the historical Arthur and the events associated with him were located in the South-West of England, then *King Arthur* masks the absence of a reality, and is level three. But the main fact about King Arthur, and Robin Hood for that matter, is that there are few facts to get in the way of theories. Perhaps then, both King Arthur and Robin Hood are destined for continual re-interpretation on the page, and on the screen; as indeed has been the case, and will, no doubt, continue to be so. As John Matthews, historical consultant to *King Arthur*, is quoted as saying, 'the legend lives on' (Fettes 2004). Somehow, the two parts of the paradox do not rest easy in bed together. Like the mixing of drinks, transparent fantasy coupled with claims of historicity lead to something of a headache.

What emerges of interest from the *King Arthur* example is, perhaps, the consequence of dallying between the world of fantasy and the world of reality. For the *Guardian*'s commentator, playing games with historicity – removing the 'sovereign difference' between the yarn and the real – erodes the magic from the one and the charm from the other. Does a meeting with the hyper-real Santa Claus at his theme-park home in Lapland compromise the 'real' Santa, that is, the Santa of the child's imagination? What about the various Santas encountered in retail premises in the run-up to Christmas? At what stage does the child, confronted with multiple Santas, begin to suspect there is no territory behind the map? These are posed as rhetorical questions.

Titanic

Titanic (1997), the third cinematic example, is potentially the most intriguing. Like *Sky Captain*, it is a nostalgia film, albeit of a very different character, but has also been described as a 'heritage film' (Terry-Chandler 2000, 68). Its charm resides in a fictional love story, which sees the voyage as a vehicle that brings together a young man and woman from otherwise separate social worlds. The love story alone occupies the first hour or so of the film. Not surprisingly, the film, in excess of three hours in length, is significantly longer than the time taken for the real Titanic to sink. The portrayal of the film on the cover of *Newsweek* some weeks after its release shows the protagonists, Jack and Rose, played respectively by Leonardo DiCaprio and Kate Winslet, holding each other, with the ship 'barely visible' in the background (Krämer 1998, 599). The actual sinking of the ship, in contrast with some previous cinematic treatments of the story, is almost a sub-plot. Within weeks of its release, *Titanic* was within sight of earning US$ 1 billion at the box office, and attracted cinema and video earnings that were historically high (Krämer 1998, 599; Terry-Chandler 2000, 67). The reasons for this are no doubt numerous, but for the purpose of this chapter, it is interesting to note that in the film, the fantasy and the real were kept resolutely separate. This, arguably, serves to create the charm for which the film has earned its reputation. First, there is the question of creating authenticity:

> Cameron's problem was to create an aura of 'pastness' so persuasive that viewers would experience not only the liner, they would experience the past itself, and he thought he achieved this by providing a tangible location, a 'there' in time and space, and transporting (the pun is particularly relevant) his viewers to it (Middleton and Woods 2001, 507).

Cameron's personal commitment to authenticity involved efforts to avoid the kind of liberties taken in the earlier examples. He is quoted as being confident that his film comes as near as possible to an experience of being in a time machine and visiting the ship, while being aware that a recreation of a past event can never be more than respectful approximation. But the love-story element of the film, though played in costume, portrays the protagonists as recognizably modern characters (Terry-Chandler 2000, 72–5). This, in its way, is simple representation, but a representation of the present in the past. The director's aspiration was that the 'heritage' half should also be as representative as possible, but on his own admission, post-hoc re-creation can only approximate to a reality that can, in any case, no longer be known. Therefore, *Titanic* is *de facto* level two. And for similar reasons, the same conclusion has to be drawn about heritage attractions, as the following discussion reveals.

Heritage Interpretation

Titanic (Revisited)

A newspaper advertisement caught the eye of the author (*Hull Daily Mail* 2004). Publicizing an award-winning touring exhibition (Clarkson 2008) at a municipal museum, it pictured Leonardo DiCaprio and Kate Winslet in character as Jack and Rose, holding each other, gazing skyward. There is no picture of a ship. The text invites us to view 'Original artefacts from the Titanic, including the dress worn by actress Kate Winslet'. Think about it. A telephone call to the museum confirmed that this was an exhibition of original artefacts from the *Titanic* (the ship, and her sister ships) and *Titanic* (the film). The exhibition thus represented a fusion of the myth and the real. There are two *Titanics*: the ship that sank and the cinematic love story. As an historic blockbuster, the film becomes a legend in its own right. Its myth rivals, perhaps eclipses, the event that served as its vehicle. The sign supersedes the real. As Baudrillard (1994, 1) puts it, 'the territory no longer precedes the map, nor does it survive it. It is nevertheless the map that precedes the territory – the precession of simulacra – that engenders the territory.'

The success of the film, coupled with the passage of time, has led to the *Titanic* (the film) being thrust into the role of a level three phase of the image, masking the absence of a basic reality. Further evidence of this shift subsists in the various attempts to attract capital to build a replica of the vessel. Such attempts have not come to fruition, but their very existence demonstrates that the word 'Titanic', as a signifier, has slipped its moorings, so to speak. No longer is 'death by drowning' the connotation signified by this word, but 'romance and opulence'. However, for those who survived the tragedy as young children, 'Titanic' can never be a floating signifier: it has only one meaning. 'I would never set foot in a new ship', said a survivor (McDonald 2000).

Beamish

Beamish is an open-air museum in County Durham, England. Set in a site of 300 acres, it comprises a drift mine, a farm and a manor indigenous to the site, as well as an assemblage of other buildings and artefacts – whose dates and provenances are made clear – imported from other locations in the north-east of England. Its theme is the transformation of life from the rural to the industrial, spanning the period from the late 1700s to the outbreak of the First World War. Costumed interpreters are employed to augment the visitor experience (Bennett 1988; Iles 1998).

Bennett (1988) and Iles (1998) share the view that Beamish projects an air of sentiment and nostalgia. Iles (1998) observes that buildings are in an improbably good state of repair, the female interpreters, as colliery wives, show no sign of hardship and outside toilets offer no olfactory indication of their presence. Bennett (1988) writes of the projection of 'a people without a politics', stating that the museum is amnesic in its avoidance of the class-based conflicts which punctuated the periods it presents. The emphasis is on an idyll, when communities were communal rather than cellular, and nostalgia can be evoked for the *Gemeinschaft* condition of working-class life (Rojek 1993, 284), rather than the privations of the time. The hardships, in particular the struggles between capital and labour, were identified by Dicks (2008) in her study of mineworkers re-employed as guides in collieries re-opened as heritage attractions. Dicks (2008, 449) argues that '... the current tour guide finds himself sandwiched between two realities, one past, one present... both the mine-worker and the tour guide have to negotiate the hidden injuries of class, and in the person of the ex-miner tour guide they come together.' I can relate, anecdotally, reactions of students to the experience of Beamish in classroom discussions. The adjective typically chosen by students who have visited the site is 'authentic'. But if then asked whether, during their visit, they picked up head lice, fleas, or saw any signs of tuberculosis in the countenances of the interpretive performers, the students were apt to modify their answer. It is thus difficult to think of any way in which interpretation of heritage can be truly representative.

It should be remembered that a day out should be fun, and that in choosing to portray the positives of past life, Beamish avoids the clichés and parodies of an 'It's grim up North' narrative. What is interesting, however, is that the students' dominant memory is of a faithful representation. The 'moment's reflection' is required to perceive the inauthenticities. The representation is faithful inasmuch as it does not seek to modify the past, even if its portrayal is selective. But, like any representation of its kind, the demands of the present and of interpretation require adaptation to the present and the modalities of heritage management and visitor attractions. Beamish is a level two phase of the image, because it 'perverts' a basic reality.

There is a sense in which the more a museum of this kind is successful in faithfully representing past times, the more those features that deviate from the real are concealed. First, few visitors will possess the cultural literacy to identify those deviations, just as viewers of televised historical costume dramas will for the most part be unable to identify the liberties that are being taken with replications of the costumes (Macdonald 2000, 110). Eco wrote, regarding the full-size replica of the Oval Office in the Lyndon B. Johnson Memorial Museum:

> The 'completely real' becomes identified with the 'completely fake'. Absolute unreality is offered as a real presence ... the aim of the reconstructed Oval Office is to supply a 'sign' that will then be forgotten as such: the sign aims to be the thing, to abolish the distinction of the reference ... (Eco 1987, 7).

At Beamish, the substance is 'real' buildings rather than reproductions; but this, if anything, intensifies the hyper-real. The closer the simulation comes to the real, the more absolute the fakery. Schouten (1998, 73–4), on the subject of theme parks, writes of '... an improved past that never existed, a clean version of a non-problematic history ... There is an apparent paradox in heritage: the closer we want to come to reality, the more we have to violate it.'

The Jorvik Viking Centre

This visitor attraction is located in the City of York, England. Owned and operated by a company established by the York Archaeological Trust, Jorvik opened in the early 1980s as an early example of a new breed of heritage attraction. Neither museum nor theme park, Jorvik has elements of both. Constructed on the site of an archaeological excavation that uncovered material remains from York under Scandinavian rule, visitors are entertained to a journey through a community reconstructed in accordance with the archaeological evidence (Meethan 1996, 229–331). In the words of the Trust's then director, Peter Addyman (1984, 45), Jorvik 'presents an immensely complicated story in an immediately accessible form'. There are no actors, but models equipped with recordings of voices using the language of the time, reconstructed using the best efforts of archaeolinguistics. After the journey, there are preserved remains of the buildings found during the excavation; and an interpretive studio where an employee, costumed but not 'in character', speaks to visitors about the way in which most recent advances in forensic archaeology can be used to reconstruct the lives and deaths of individuals from their human remains.

Jorvik contrasts with Beamish in that it is for the most part an exhibition, using theme park techniques to convey its message. Unlike Beamish, its 'Viking' toilets smell, although far less so than the conditions of the age might have engendered. Indeed, as one commentator suggests, the simulated smells could easily be mistaken for that of operating machinery (Silver 1988, 189). The barely concealed antagonism of another commentator (Schadla-Hall 1984) resonates with the disquiet with which the growth of heritage centres in the 1980s had been greeted in the traditional museums world. As an example of the hyper-real, Jorvik arguably lies somewhere between level two and level three. While posing as an authentic recreation, it is necessarily sanitized in the same way as Beamish in order to be acceptable for present-day visitation. By this assessment, it is level two. It can also be argued that Jorvik is close to level three, because, unlike Beamish, Jorvik's 'reality' has not been known or recorded in anything but sparse historical and archaeological evidence. Although one does not doubt the conviction of the then director of the York Archaeological Trust, that the exhibition is 'reasonably close to the truth' (Addyman 1984, 45), it is impossible to know the extent to which the display coincides with what was real. In this sense, Jorvik is perhaps level three. The 'basic reality' behind the exhibition can never be known and is therefore absent.

Conclusion

Emerging from this study, there are four particular points to note. First, the hyper-real operates at different levels of intensity, from pure representation of the real, through to the pure simulacrum where the original is absent or superseded by its own sign. It has been shown that Baudrillard's four phases of the image can be deployed as an analytic model in diagnosing different intensities of the hyper-real in the representation of the past. Quirky though his writings may appear, there is a seriousness at their centre which can only be conveyed in his unconventional way. In the assessment of a journalist (Leith 1998, 14), '... Baudrillard's work contains a compelling dry humour. Perched on the dangerous ledge between the completely bonkers and the scarily lucid, he keeps prodding you, through humour, with the idea that, in a way, he might be right'.

Second, there is a paradoxical pursuit of the authentic. The proliferation of information through the media has led to a demand for images and styles to fill up those media. 'The past' has become a supermarket of styles to be raided and reproduced. A heritage industry has emerged to perform this service for us. Interest in the past revolves, unsurprisingly, around 'what it was like'. Paradoxically therefore, the authenticity of the object, genuine or otherwise, is a *de rigeur* claim and has been replaced by the authenticity of the viewer's engagement with whatever is represented, the latter being a key aim of the heritage entrepreneur. The meaning of 'authenticity' is elusive; Rojek terms the question 'chronically insoluble' (1993, 285), especially when it is juxtaposed with 'reality', as I noticed when I spoke at a conference in the Netherlands and made passing reference to authenticity. At lunchtime, around a small table with an attractive posy of flowers in the middle, delegates demanded to know more. 'There is a difference between the real and the authentic', I said. 'Consider these flowers: they are not real, but they are authentic'. Mild outrage ensued. This was Holland. The flowers were real. Fake flowers were anathema in the country which prides itself on being the purveyor of blooms to the civilized world. The flowers were in such good condition I was fooled into thinking they were artificial: more flower than flower and, hence, fake. But to describe them as authentic seemed to present no problem, and that sheds light on the meaning of the word. The authentic is not the same as the real. The authentic may be real, but the principal meaning of the word is to do with faithful representation and accuracy. So a piece of Baroque music performed on authentic instruments can mean two things: either that the instruments date from the Baroque period, or that they are truthful reproductions; or, as is perhaps more likely, that they are a mixture. In either case, the authenticity of the object is replaced by the authenticity of the listener's experience, which, as with heritage interpretation and historical movies, is often the only reality left when the simulacra are exposed and historicity is eroded.

The third point to note is that in the drive to reproduce the past in a suitably vivid way, historicity is eroded. In heritage interpretation, the more layers of interpretation are added – one could say, the greater the attempt to produce 'realism' – the greater the departure from the known and the real. The Chorus of Shakespeare's *Henry V* may encourage his audience, at the start of the play, to 'Piece out our imperfections with your thoughts'; but the contemporary drive is toward spectacle and sensation (Rojek 1993, 285) and a concomitant demand on the consumer for passivity, rather than active use of imagination (Finn 2002). In cinematic film, not surprisingly, the requirement to entertain is paramount and liberties with the past more likely to be taken, as in the examples given.

The fourth point is that when the boundary between fantasy and the real are blurred, this is somehow to the detriment of both. The magic of the one, and the charm of the other, are compromised. *The Guardian*'s reaction to *King Arthur* is an example of that. Perhaps another example is that of the entertainer Madonna. She started her career in the mid-1980s as a Marilyn Monroe look-alike, and has since passed through many reinventions. Marilyn Monroe was an actress called Norma Jean; and there are testimonies that Norma could turn 'Marilyn' on and off at will (MacCannell 1992, 234–5). The distinction between the woman, and the character, was clear. Perhaps the enduring charm evoked by Marilyn is due to this distinction. Can it be said about Madonna, that there is a distinction between her stage re-inventions and the 'real' woman? One of her hits was titled, 'Who's that girl?' That, perhaps, is *la belle question.* If Madonna is to be admired, it is as an entertainer, a survivor, and her ability to change her act through reinvention. Whether 'charm' can be ascribed to her is doubtful. I write subjectively. No doubt there will be those who disagree.

Epilogue: Memory Plays Tricks

As a final thought, it would be useful to bear in mind that, if visualizations of the past are selective and distort the past, *so does the human memory*. Larsen's discussion of the psychology of tourist experience leads him to conclude that remembered tourist experiences are matters strong enough to have entered the subject's long-term memory. The content of the long-term memory consists of flash-bulb memories, vivid in nature. Citing evidence, in particular studies of memories of the 9/11 attacks in New York, he shows that flashbulb memories, like everyday memories, become inaccurate with the passage of time (2007).

This resonates with the point made above, regarding the shift of emphasis from authenticity of object to authenticity of experience (see Crouch this volume). The insight offered by post-structuralism is that meaning is authored in the mind; that which is encountered is merely the stimulus, and thus each individual can be expected to construct a unique meaning from the same stimulus (Derrida 1998, 158; Walsh 1992, 55; Crouch this volume). If heritage interpretation, despite (or because of) best efforts at authenticity, produces a hyper-real; and if cinema, more prone to taking liberties with historicity, treats us (for example) to crossbows several hundred years before they were invented; in the end, we the consumers take that information and author individual meanings. More than that, as Larsen suggests, those meanings can be expected to mutate over time as our memories perform their own selection. What images and words mean to an individual are difficult to divine, even when their provenance is known. As a character memorably put it in the final scene of Orson Welles' *Citizen Kane* (1941), 'I don't think any word can explain a man's life'. And, as Shakespeare put it several centuries earlier in his final sole-authored play before retirement:

> We are such stuff as dreams are made of, and our little life is rounded with a sleep (*The Tempest*, IV, i).

References

Addyman, P. (1984), 'Jorvik: Rebirth of a City', *History Today* 34, 337–42.

Alcock, L. (2001), *Arthur's Britain: History and Archaeology A.D. 367–634.* (Harmondsworth: Penguin Classic History).

Baudrillard, J. (1993), *Symbolic Exchange and Death.* (London: Sage Publications).

Baudrillard, J. (1994), *Simulacra and Simulation.* (Michigan: University of Michigan Press).

Bennett, T. (1988), 'Museums and "the People"', in R. Lumley, (ed.) *The Museum Time-Machine*, 63–85 (London: Comedia).

Bern Info (2008), *Plattform für Kunst, Kultur und Gesellschaft: Graf Zeppelin in Bern* [Online]. Available at: http://www.g26.ch/texte_018.html [accessed: 2 February 2008].

Carr, E. (2001), *What is History?* (Basingstoke: Palgrave Macmillan).

Clarkson, A. (2008), *Swansea Museum Exhibition One of Best in UK* [Online]. Available at: http:// www.titanic-titanic.com/forum/viewtopic.php?t=4217 [accessed: 16 October 2008].

Constable, C. (2006), 'Baudrillard Reloaded: Interrelating Philosophy and Film via *The Matrix Trilogy*', *Screen* 47:2, 233–49.

Davies, M. (1989), 'A Loss of Vision', *Leisure Management* 9:10, 40–2.

Derrida, J. (1998), *Of Grammatology*, Corrected Edition. (Baltimore: John Hopkins Press).

Dicks, B. (2008), 'Performing the Hidden Injuries of Class in Coal-Mining Heritage', *Sociology* 42:3, 436–52.

Eco, U. (1987), *Travels in Hyperreality.* (London: Picador).

Fettes, M. (2004), 'Quest for the Truth about Arthur and Scotland', *Edinburgh Evening News*, 13 July 2004.

Finn, M. (2002), 'From Sport to Spectacle: The Emergence of Football as a Destination Product Attribute', in R. Voase (ed.) *Tourism in Western Europe: A Collection of Case Histories*, 171–91 (Oxon: CABI International).

French, P. (2004), 'Review: *Sky Captain and the World of Tomorrow*'. *The Guardian* [Online, 2 October]. Available at: http://www.guardian.co.uk/features/featurepages/0,1318350,00.html [accessed: 5 June 2008].

Hewison, R. (1987), *The Heritage Industry: Britain in a Climate of Decline.* (London: Methuen).

Hull Daily Mail (2004), 'Honour and Glory', advertisement, 12 April 2004, 21.

Iles, J. (1998), 'History as Leisure: Business and Pleasure at Beamish', in N. Ravenscroft, D. Phillips and M. Bennett (eds) *Tourism and Visitor Attractions: Leisure, Culture and Commerce*, 183–94 (Brighton: Leisure Studies Association).

Jameson, F. (1984), 'Postmodernism, or the Cultural Logic of Late Capitalism', *New Left Review* 146, 53–92.

Jameson, F. (1985), 'Postmodernism and Consumer Society', in Foster, H. (ed.) *Postmodern Culture*, 111–25 (London: Pluto Press).

Jones, J. (2004), 'Death of a Legend', *The Guardian* [Online 23, August]. Available at: http://www.film.guardian.co.uk/features/featurepages/o,,1299747,00.html [accessed: 5 June 2008].

King Arthur (dir. Antoine Fuqua, 2004).

Krämer, P. (1998), 'Women First: *'Titanic',* Action-Adventure Films and Hollywood's Female Audience', *Historical Journal of Film, Radio and Television* 18:4, 599–617.

Larsen, S. (2007), 'Aspects of a Psychology of the Tourist Experience', *Scandinavian Journal of Hospitality and Tourism* 7:1, 7–18.

Lee, C. (2005), 'Lock and Load(up): The Action Body in *The Matrix*', *Journal of Media and Cultural Studies* 19:4, 559–69.

Leith, W. (1998), 'I'm Not a Real Photographer', *Observer Life*, 15 February 1998, 12–17.

Lowenthal, D. (1998), *The Heritage Crusade and the Spoils of History.* (Cambridge: Cambridge University Press).

Lyotard, J-F. (1984), *The Postmodern Condition: A Report on Knowledge.* (Manchester: Manchester University Press).

MacCannell, D. (1992), *Empty Meeting Grounds: The Tourist Papers.* (London: Routledge).

MacDonald, H. (2000), 'Titanic Survivor Condemns Plan to Rebuild Ship as Cashing in on Tragedy'. *The Observer* [Online, 31 December]. Available at: http://guardian.co.uk/uk/2000/dec/31/theobserver.uknews2 [accessed: 3 September 2008].

Marwick, A. (1989), *The Nature of History.* (Basingstoke: Macmillan).

Meethan, K. (1996), 'Consuming (in) the Civilized City', *Annals of Tourism Research* 23:2, 322–40.

Middleton, P. and Woods, T. (2001), 'Textual Memory: The Making of *Titanic's* Literary Archive', *Textual Practice* 15:3, 517–26.

Moffat, A. (1999), *Arthur and the Lost Kingdoms.* (London: Weidenfeld and Nicolson).

Muri, A. (2003), 'Of Shit and the Soul: Tropes of Cybernetic Disembodiment in Contemporary Culture', *Body & Society* 9:3, 73–92.

Omega Watch Company (2008), *Omega Ambassadors*, [Online]. Available at: http://www.omegawatches.com [accessed: 1 September 2008].

Possession (dir. Neil LaBute, 2001).

Poster, M. (ed.) (2001), *Jean Baudrillard: Selected Writings.* (Cambridge: Polity Press).

Rojek, C. (1993), 'After Popular Culture: Hyperreality and Leisure', *Leisure Studies* 12:4, 277–89.

Schadla-Hall, R.T. (1984), 'Slightly Looted: A Review of the Jorvik Viking Centre', *Museums Journal* 84, 62–4.

Schouten, F. (1998), 'Authenticity and Real Virtuality', in S. Scraton (ed.) *Leisure, Time and Space: Meanings and Values in People's Lives*, 73–8. (Brighton: Leisure Studies Association).

Silver, J. (1988), '"Astonished and Somewhat Terrified": The Presentation and Development of Aural Culture', in R. Lumley (ed.) *The Museum Time-Machine*, 170–95. (London: Comedia).

Skene, W. (1988), *Arthur and the Britons in Wales and Scotland*, Bryce, D. (ed.) (Lampeter: Llanerch).

Sky Captain and the World of Tomorrow (dir. Kerry Conran, 2004).

Stuart-Glennie, J. (1869), *Arthurian Localities: Their Historical Origin, Chief Country and Fingalian Relations.* (Edinburgh: Edmonston and Douglas).

Terry-Chandler, F. (2000), 'Vanished Circumstance: *Titanic*, Heritage, and Film', *International Journal of Heritage Studies* 6:1, 67–76.

The Matrix (dir. Wachowski Brothers, 1999).

Titanic (dir. James Cameron, 1997).

Walsh, K. (1992), *The Representation of the Past: Museums and Heritage in the Post-Modern World.* (London: Routledge).

Yorkshire Robin Hood Society (2002), *Yorkshire???* [Online]. Available at: http://www.robinhoodyorkshire.co.uk [accessed: 3 September 2008].

PART III
Visual Culture and Heritage Tourism

Chapter 8
'Wild On' the Beach: Discourses of Desire, Sexuality and Liminality

Annette Pritchard and Nigel Morgan

This chapter examines television representations of the beach as an embodied site of transgression framed by discourses of liminality and the carnivalesque. In the following pages we will suggest that the peculiar configurations of open and negotiable abstracted spaces of the contemporary resort beach are represented as offering a range of opportunities for display, performance and transgressive behaviour. The beach is a space of 'natural heritage' (Boyd and Butler 2000) and has been conceptualized as a liminal, in between place, neither land nor sea, where the normal social conventions need not apply. Much of the work underpinning this conceptualization of the beach is outside mainstream tourism studies and can be traced to sociology (Shields 1991), social history (see Walton 2000) and cultural geography (see Preston-Whyte 2004), scholarship which in turn builds on the work of van Gennep (1960) and Turner (1974). Central to the notion of liminality is its transitory, betwixt nature, whether manifested in terms of social life, space or time, so that '[i]n this gap between ordered worlds almost anything may happen' (Turner 1974, 13). By extension, this conceptualization of the beach has similarly constructed the historic seaside resort as a ludic and unconventional site. Thus, Shields (1991, 75) has explored the liminal, carnivalesque and illicit pleasures associated with the British seaside resort of Brighton, particularly in the 1930s and 1950s, which became 'the topos of a set of connected discourses on pleasure and pleasurable activities ... without which our entire conception and sense of a beach would be without meaning'.

However, whilst Shields (1991, 73) argued that today many Western consumers no longer need 'to create marginal zones, such as the seaside was, for reckless enjoyment', we argue that contemporary beach resorts remain very much at the heart of such discourses. In fact, we will suggest below that the transgressive behaviour witnessed in them shares much with the outpourings of excess and the challenging of norms which was historically found in the seaside carnival, described by Shields (1991, 94) as 'a mark of resistant bodies which at least temporarily escape or exceed moral propriety'. Of course, such behaviour is not new and the conducting and establishing of new sexual relationships away from the norms and values of the everyday whilst on holiday has a long tradition, from the eighteenth-century scandals in Bath, 'the first of Britain's large specialized resorts to pull in a national visiting public' (Walton 2000, 5) to today's holiday resorts. Indeed, Ryan and Hall (2001) note that surveys suggest that somewhere between just under a tenth and just under a quarter of holidaymakers admit to having sex with a new partner whilst on holiday.

This transgressive behaviour itself reflects the increasingly overt and explicit emphasis on sexuality in many forms of Western popular culture, which has disguised a more ambiguous and open-ended search for pleasure and self-expression focusing on the body (Savage 1990); a sexualization of the public sphere which has also meant a greater emphasis on embodiment, nudity and sexuality. Attwood (2006) has recently examined our sexualized culture and contemporary preoccupations with sexual values, practices and identities as well as the emergence of new forms of sexual experience, pornographication, taste formations, post-modern sex and intimacy and sexual citizenship. In this movement, the beach is perhaps *the* site where we are invited to 'Get undressed – but be slim, good looking, tanned' (Foucault 1982, 57). Thus, we can say that the beach is the archetypical territory of desire that embodies such processes of seduction, impression management, self-expression, and the construction, exchange, and interpretation of embodied sign-values. It is dominated by a particular form of masculinity and femininity based on the athleticism and beauty that is highly prized in contemporary society and replicated in countless media images of the perfect suntanned beach body.

This chapter will locate this sexualization of the public sphere and the liminality of the beach in a wider discussion of embodied tourist experiences and examine representations of the (un)dressed beach body in the television series *Wild On* and *Naked Wild On 2*. Based on a textual analysis of these programmes, filmed largely in Florida and the Caribbean, the chapter has two goals. First, it seeks to develop our work on tourism's gendered visual rhetoric (Pritchard and Morgan 2000a,b, 2005, 2007) by premising that such programmes form another element in the circle of representation (Hall 1997) and contribute to the stereotypical representation of men and women. The circle of representation or the circuit of culture is a concept which is established across a range of research fields, although it is presented slightly differently in each. The basic idea is that certain images circulate within a culture and take on particular meanings, associations and values. This conceptualization recognizes that language, representation and meaning are connected in a continuous circle so that a set of discourses – by which we mean frameworks which embrace particular combinations of narratives, concepts and ideologies – become so powerful that, reinforced over time, they come to form a closed self-perpetuating system of illusion or a 'way of seeing' the world (see Morgan and Pritchard 1998). In this chapter, we are particularly concerned to analyse the connections between tourism representation and the body through a form of cultural criticism, interrogating the capacity of the travel-related entertainment media to perpetuate overtly sexual and sexist ideological constructs. We suggest that, just like other cultural forms, these discourses of embodied desire, sensuality and hedonism frame the tourist experience. Second, the chapter seeks to illustrate the power of travel programmes as agents of cultural pedagogy and body discipline, and argue that such programmes objectify women and indirectly contribute to body image anxieties. We begin, therefore, with a brief review of the complex connections between the body, the beach and tourism. This is followed by our analysis of several programmes from the two television series, along with our conclusions.

The Body Beautiful

Until very recently, 'the body' had been absent from tourism enquiry, reflecting the field's masculinist, disembodied research traditions (although see Selwyn in this volume for a recent analysis of the body within tourism). Although the body occupies a central place in the tourist imagination, and even though the tourism industry is directly concerned with global circuits of bodies, body images and body servicing, the body itself remains taken-for-granted and under-theorized by tourism scholars. Indeed, Eeva Jokinen and Soile Veijola (1994) have argued that the overwhelming emphasis on the tourist gaze has under-served the embodied politics of travel and sexuality, whilst Jennie Small (2007) goes further and argues that academic snobbery has led scholars to marginalize the corporeally-oriented 'beach' tourist in favour of the mind-oriented 'cultural' or 'heritage' tourist. But we are slowly seeing a shift in focus as the field starts to address the new work on identity, difference, the body, gender and post-structural theories of language and subjectivity that has forced such a rethinking in the wider social sciences. As a result, tourism scholars have now conceptualized, re-theorized and incorporated concepts and themes such as sensuousness and embodiment, performativity, the senses, and materialities and mobilities (see Pritchard et al. 2007 for a review).

In this chapter, we focus specifically on travel entertainment programmes as products of cultural discourses and inscriptions, and analyse such television programmes as a discursive domain which sets parameters around the presentation of particular social and cultural bodies. We also explore the significance of how the sexualized female body is constructed in the social world. The body has been termed 'the vehicle *par excellence* for the modern individual to achieve a glamorous lifestyle' (Davis 1997, 2, italics in original), and in this chapter we explore those subtle but pervasive processes of discipline and normalization of the male and female body that are communicated through cultural representations of tourism. Travel has been described as having an intimate connection with sexual desire and scopophillia (Mackie 2000) and recent analyses have examined the gendered and sexed dimensions of the tourist gaze, exposing the often implicit masculine possessor of this gaze (see Pritchard and Morgan 2005, 2007). Whilst such analysis has noted the prevalence of sexualized representations of women in tourism marketing literature, there has been less analysis of how 'tourist' bodies are seen materially and symbolically in wider popular cultural forms such as television programmes.

Since the media industries are profound and unsuspected agents of cultural pedagogy (Williamson 1986; Tomlinson 1991), the relationships between the media, the body and tourism deserve more attention if we are to understand the ways in which sensual pleasures and the fulfilment of bodily desires are constructed as part of the tourism experience. In his revision to *The Tourist Gaze*, John Urry (2002) recognizes that television travel programmes play an important role in the discursive formation of the gaze and in the consumption of place. Moreover, the significance of such representations cannot be underestimated because they play a part in shaping our own embodied identities and practices, as:

> The symbolic content of representations is internalized as knowledge that later shapes the production of gender performance ... there is a clear relationship between the discursive and the non-discursive ... texts ... impact on physical behaviours and display and vice versa, in a recursive loop of meaning (Knox and Hannam 2007, 271).

Certainly, contemporary media is saturated with body representations and body stories; our culture prizes young, slim and beautiful bodies, whilst simultaneously problematising and marginalizing ageing, overweight and disabled bodies. Our bodies are now objects for display (Featherstone 1991) and they are increasingly central to, and have come to symbolize, our sense of self and identity as people define their bodies as '*individual* possessions which are integrally related to their sense of self identities' (Shilling 2003, 28, emphasis in original). From an ever younger age, our bodies are in a constant state of becoming and we increasingly try to manage and maintain them, immersing ourselves in body projects through diet, fitness regimes, body building and sculpting, health plans, cosmetics and plastic surgery. Indeed, body modification techniques such as plastic surgery, and non-surgical procedures (such as botox and skin peels), are visible evidence of 'a body nostalgia' whereby people try to stop or turn back the clock (Gullelte 1997). Yet, as our ability to exert control over our bodies has grown, so have our body anxieties become ever more acute. Western women are particularly vulnerable to what Chernin (1983) terms 'a slenderness tyranny' and are subject to a barrage of images of thin, young, air-brushed female bodies. In the emotional domain, sexualization and objectification undermine confidence in, and comfort with, one's own body, leading to a host of negative emotional consequences such as shame, anxiety, and even self-disgust (Slater and Tiggemann 2002). Today, society is increasingly concerned with obesity and its associated health consequences and the pursuit of the body beautiful is big business, with attempts to delay the physical and aesthetic signs of ageing fast developing into an obsession. There is a proliferation of 'health' and cosmetic products claiming to defy signs of ageing, and in North America the number of non-invasive cosmetic procedures increased by almost 800 per cent between 1997 and 2005, so that over US$ 12.5 billion is now spent there annually on such cosmetic treatments to counteract physical signs of age (Esfahani 2006).

As individuals we are therefore able to exert more and more control over our bodies, but at the same time, ever greater numbers of us feel constrained by our overweight or physically imperfect bodies that do not conform to some unattainable beach babe or muscle beach image. And here we turn to the heart of this chapter – visual representations of the sexualized beach body. The beach as a site of tourism is dominated by a particular form of masculinity and femininity based on athleticism and beauty, and as a space of bodily undress it thus becomes an anxious place of display and body management, particularly amongst women. In 'beach cultures' there is a greater emphasis on needing to spend time at the beach and to be tanned as a result of greater bodily exposure and public scrutiny; for instance, a state-wide study of New South Wales adolescents' health concerns found that 'body image' was a concern for twice as many adolescents in beach and coastal locations as in inland locations (Quine et al. 2003). Those of us who do not live near the coast might prepare for a beach holiday through bodily regulation or deprivation (Jordan 2007) and buy ourselves the latest swimsuit specifically for the trip (Westwood 2004; Ambrovici 2007). Women's magazines regularly offer diet programmes that individuals follow in pursuit of a 'celebrity bikini body', and television programmes such as *Inch Loss Island* transpose the diet experience to a tourist resort – directly linking the latter to discourses of bodily regulation and control. This is an ideology of bodily management and objectification in which the tourism industry is thus heavily implicated – both in its symbolic and material realms.

Tourism also plays a central role in the overt and explicit emphasis on sexuality in many forms of popular culture, which Savage (1990) has suggested is linked to a contemporary search for pleasure and self-expression focusing on the body. This sexualisation of the public domain has brought with it a concomitant emphasis on nudity and sexuality, and here again, as we will see, tourism practices are implicated. As other writers have documented (for example Ryan and Hall 2001), tourism offers many opportunities for bodily encounters and sexual behaviours which might not be possible in our home environments, a situation clearly exposed in the many recent docu-soaps on UK television such as *Ibiza Uncovered, Holidays Reps* and *Pleasure Island.* These types of programmes very clearly demonstrate the liminal, marginal nature of many tourism environments as they set out to expose extremes of behaviour (in terms of drinking and sexual encounters) and to titillate the (assumed) male viewer by offering heavily sexualized images of women.

There are four key components to sexualisation, namely: a person's value comes only from his or her sexual appeal or behaviour, to the exclusion of other characteristics; a person is held to a standard that equates physical attractiveness (narrowly defined) with being sexy; a person is sexually objectified – that is, made into a thing for others' sexual use, rather than seen as a person with the capacity for independent action and decision making; and/or sexuality is inappropriately imposed upon a person. All four conditions need not be present; any one is an indication of sexualization and anyone can be sexualized, although it is most likely to be imposed on women and children (American Psychology Association Task Force on the Sexualization of Girls 2007). Alarmingly, some 40 years after the birth of the Western feminist movement, study after study continues to demonstrate that virtually every media form provides evidence of the sexualisation of women, including television (Grauerholz and King 1997), music videos and music lyrics (Vincent 1989; Gow 1996), movies, magazines (Krassas et al. 2001, 2003), sports media, video games, the internet and advertising (Lin 1997). In these and many other studies, findings indicate that women, more often than men, are portrayed in a sexual manner (for example, dressed in revealing clothes, with bodily postures or facial expressions that imply sexual readiness) and are objectified (for example, used as a decorative object or as body parts rather than a whole person). Moreover, a narrow and unrealistic standard of physical beauty is heavily emphasized across these media forms.

These are the models of femininity presented to adults and children in Western society and frequent exposure to media images that sexualize girls and women very clearly affects how girls conceptualize femininity and sexuality. Girls and young women who more frequently consume or engage with mainstream media content offer stronger endorsement of sexual stereotypes that depict women as sexual objects (Ward and Rivadeneyra 1999; Ward 2002; Zurbriggen and Morgan 2006).They also place appearance and physical attractiveness at the centre of women's value, and psychological researchers have identified *self-objectification* as a key process whereby girls learn to think of and treat their own bodies as objects of others' desires (McKinley and Hyde 1996; Frederickson and Roberts 1997). In self-objectification, girls internalize an observer's perspective on their physical selves and learn to treat themselves as objects to be looked at and evaluated for their appearance, and numerous studies have documented the presence of self-objectification in women more than in men. However, such is the extent of our sexualizing society that recent research has also highlighted how body image issues are affecting boys and men. In fact, researchers and the media have recently drawn attention to an apparent crisis in contemporary forms of masculinity, marked by uncertainities over men's social role and identity, sexuality, work and personal relationships (Frosh et al. 2003). Thus men's magazines, for example, are caught between attempts to construct masculinity as a form of certitude, while at the same time responding to a world of changing gender relations (Jackson et al. 2001).

From this brief review of some of the connections between tourism, the media and body image, we can see how the beach as a space of bodily undress is an uncomfortable zone characterized by bodily display and sexualized performance. In 'beach cultures', there is a greater emphasis on body management as a result of greater bodily exposure and public scrutiny, and the beach itself is dominated by a particular form of masculinity and femininity based on athleticism and beauty. In addition, having contextualized the increasing sexualisation of today's society, we have also outlined how virtually all media forms sexualize women and girls. The next part of the chapter illustrates how certain television travel programmes contribute to this sexualization of the female body and to the denigration of 'unattractive' bodies who fail to conform to an unattainable ideal of beauty.

Consuming Bikini Beach Babes

Wild On! was a travel show produced from 1997 until 2003 by the E! Entertainment Channel. The programme, and its successor *Naked Wild On 2*, offers a mix of staged segments and 'fly-on-the-wall' filming of tourists in vacation resorts, and is an example of the complex interweaving of entertainment, capital, identity, mobility and image that is the dominant feature of contemporary society and culture. Their presenters are all female and include the former fashion and Playboy models Jules Asner, Cindy Taylor, Brooke Burke, Victoria Silvstedt, Karen McDougal and Ashley Massaro – examples of hyper-sexualized women who embody the programme's fusion of entertainment, travel, fashion and sexuality. These female hosts are usually featured engaging in performances and activities where their bodies are prominently displayed – dancing in clubs, enjoying active sports and, above all, appearing on beautiful beaches in revealing swimsuits. The E! broadcasts usually censored nudity, however the *Naked Wild On 2* series is uncensored and much more voyeuristic; the female presenters have been replaced by a male narrator who is disembodied, simply passing comment on the bodily performances and appearances of others – an example of the separation of (the male) mind and (the female) body. Whereas the female hosts embodied the previous show as active (and sexually attractive) participants, this male narrator is now a distanced spectator and critic, mediating the carnivalesque and grotesque for the viewer's titillation and amusement. But in this television series, it is the female body which is most definitely *the* focus of the camera – whether partially clothed or naked, shots of women and female body parts (particularly breasts and bottoms) are accompanied by pulsating dance tracks and the male voiceover whose comments are filled with innuendo and double entendres.

Contemporary society continues to ascribe greater cultural capital to those who display evidence of conventional attractiveness and athleticism in their bodily practices, and these programmes show female bodies that epitomize hyper-femininity and erotic beauty. In fact, the whole tone of the series echoes the so-called 'lad culture' of recent years, where women are portrayed as objects to titillate and gratify the male voyeur in carnivalesque sites of seduction (Ryan and Hall 2001), whether in wet T-shirt competitions, posed in swimsuits, gyrating to dance beats or portrayed in staged lesbian clinches. With the shedding of clothes comes public licence to display the sexual body for the sexualized gaze of others, be they fellow tourists or, as in this case, television viewers. When we watch this programme, we enter a male playground, a liminal zone of hedonism where, as the narrator tells us, '... surf is not the only thing up in this sea of oiled bodies and bronzed flesh'. In this television show, as we shall see, certain bodily norms are valorised and celebrated: desirable beach bodies are white but bronzed, young, smooth and hard, pert breasts and tight bottoms are de rigor. Indeed, many of the women featured are reduced to these particular body parts. In today's society, our bodies are bearers of symbolic values and some are 'worth' much less than others – certain bodies are out-of-place in our body conscious, highly sexualized world, thus in the beach cultures of these shows ageing and over-weight women provide figures of fun or disgust, whilst men's bodies are seen to be unremarkable and are unremarked upon, seemingly immune to the social scrutiny of the narrator and his audience.

In one episode (*Bikini Beach Babes*), the viewer is invited to '... travel the globe and meet the world's most beautiful bikini beach babes' on beaches from Italy and Hawaii to Daytona Beach. These beaches are full of 'fit girls in bikinis' and the male narrator is keen to point out that the only thing 'we' like better is fit girls 'without the bikinis'. In the segment shot in the USA, the focus is initially on Florida's 'spring break' season when girls are 'easy and available'. Spring break is considered a rite of passage amongst North American college and university students aged between 19 and 23, and 1.5 million of them travel to popular travel destinations between early March and the beginning of April to, in contemporary language, 'break loose and party'. Currently, the most popular destinations include: Panama City, Daytona, South Beach, Key West (all in Florida), Lake Havasu City (Arizona), South Padre Island (Texas), Paradise Island and Negril (both in Jamaica) and Acapulco and Cancun (both in Mexico, where the legal drinking age of 18 is the main attraction). According to the US Department of State, over 100,000 American teenagers travel to Cancun alone over spring break each year. In addition, there are numerous cruises that depart during spring break each year.

These spring break trips are a contemporary form of carnival (Bahktin 1984), a liminal transient space based on a morality of sexual and substance excess and spectacle. Just as medieval festivals harked back to pleasurable, unrestricted and uncontrolled eras through their celebrations of gluttony, excess and the grotesque, these events (and television shows like *Wild On!* that feature them) are framed by discourses of rampant sexism, questionable sexual health practices and an ideology of sexual gratification. The events themselves are sites of sensory overload and symbolic bombardment as tobacco, alcoholic and soft drinks manufacturers, bars, clubs, student tour companies, credit card companies, the destinations themselves and sunscreen manufacturers sponsor parties, music concerts, and drinking and dancing contests. Wet T-shirt contests are typically staged poolside in arenas dominated by sponsors' hoardings and banners, and playboy playmates can be seen judging bikini contests in a wrestling ring sponsored by *playboy.com* and surrounded by a crowd of men howling, clapping and taking photographs. The competition amongst the male audience to watch and touch the women reinforces the notion that their bodies are ornamental trophies or public property subject to sexual scrutiny (Rojek and Urry 1997), where '"to be" for women is "to be seen" ... by the masculine eye or vision' (Ambramovici 2007, 109). Symbolic, communal, personal and corporate imagery merge as semi-clad male students and scantily-clad, highly sexualized Hawaiian Tropic girls become entwined in simulated sex acts. These women, all beautiful, bronzed and wearing skimpy red swimwear, are dehumanized and commodified as they pose for pictures, hand out and spray people with free samples of the sunscreen in this fiesta of noise, colour and sex. They suggestively rub against each other, applying the sun cream to themselves and numerous men, as the narrator comments: '... whilst sun cream is a good idea to protect skin on Spring Break, it's just another excuse to feel up a person you've never met before'. These women also host and judge a variety of beach games and events, such as kissing and clothes swapping contests and belly dancing competitions.

Mikhail Bahktin, the Russian sociologist and literary critic, analysed sixteenth-century feudal festivals, highlighting how the peasants sought to break free (at least temporarily) from the restraints of the church through humour, performance, food and various bodily delights:

> Carnivals were times and places of inversion, sanctioned deviance and reversals of norms. It stood opposed to the official feasts and tournaments that celebrated the power of the elites, who were instead parodied, mocked, hectored, and ridiculed. Moral boundaries from the political to the erotic were transgressed (Langman 2003, 226).

Feudal carnivals represented a temporary and unmediated opportunity for vice, sensations and excess. As a parallel celebration of excess, exotic imagery, fantasy, sensation and sexuality, the spring break can be seen to have similarities with the carnivals of old. Spring break seems to be traditionally a time for experiment and for subverting norms; one of its features seems to be sexual experiment, and lesbian experiences feature highly in the popular imagination of the typical spring break. For example, a major online commercial photography bank (Istockphoto.com) offers images of conventionally attractive women kissing in erotic poses, entitled 'spring break' and 'what did you do on spring break?' This very much echoes the transformation of sexual norms during carnival week in Brazil (also regularly featured in the *Wild On!* series). Here, the cultural scripts of carnival also actively encourage the loosening of sexual restraint and the pursuit of sexual pleasures. Carnival becomes a time for losing mastery over one's body and merging with the bodies of the crowd. Costumes are worn that emphasize sexual body parts, public nudity and semi-nudity are commonly accepted and there is much erotic dancing and drinking of alcohol (Parker 1991). Such festivals are a concoction of sound, colour, heat and sweat: a spectacle, a festival of sensory overload. But, unlike the feudal festivals, as Langman (2003) points out in her discussion of contemporary forms of the carnivalesque such as the Rio de Janeiro Carnival, in the global age of symbolic capitalism and sign-consumption, these festivals of extravagance are produced and mediated by corporate organizations. Whilst retaining a lessening or subversion of social norms, contemporary carnivals such as spring break lack the spontaneity and authenticity of the historical events. Yet, whilst they construct a temporary forum for indulgence and excess, it could be argued that although such fantasies are an escape from the confinements of contemporary society, they are also an illustration of the sexualisation of society itself and an emerging status quo of hyper-sexualized images of women. Thus, in *Naked Wild On! 2: Mardi Gras*, the male narrator informs the viewer that at carnival '... you're guaranteed to get away with stuff that would usually land you in trouble with the law' and that during Mardi Gras people take to the streets, 'get naked' and the girls 'show their tits' in exchange for beads.

In *Wild On! Spring Break: From Class to Crass*, the viewer follows a group of US college students' journey into a world of 'drunken debauchery, shameless nudity and downright stupidity'. In Panama City Nevade (also known as Red Neck Riviera), it's '... all stars, bars and girls with no bras', where the narrator tells us '... people just want to let live, they're sick of studying ... we just want to party on [doing] ... stuff you normally wouldn't do [including] get wasted [and] naked ... with the drink flowing freely, the mischief soon follows ... who needs self respect when you've got drink'. Here, bodily matters and consequences are ever present – partying, excess alcohol, vomiting, public sex acts and beach games (such as 'snogging' contests, belly dancing, swimsuit swapping contests, water-assisted bikini strip contests and wet T-shirt competitions for the women and belly flop contests for the men) are all encouraged as young people are released from the everyday social norms and conventions which govern behaviour.

Away from carnival and spring break settings, the discourse of sexuality remains the same in these television shows. In *Naked Wild On! 2: Bump and Grind*, the focus of the programme are the 'hot body contests' in South Beach Miami. In this resort, the narrator tells us '... you've got to head for the pool [which is] ... shallow, dirty and filled to bursting with horny men'. Here, women compete in erotic dancing contests and there are numerous camera shots of breasts and bottoms and simulated lesbian sex acts. There is even a glass 'sex box', where men wearing rubber gloves line up to touch the women, whose bodies have become public space, fetishized and fragmented, a corporeal touchscape 'burdened with men's meanings and interpreted through masculinist discourse' (Pritchard and Morgan 2005, 55). In another programme in the series, *Wild On! Bikini Beach Babes*, we're told that in Daytona Beach (famous for its 23-mile long beach and racetrack) the only thing 'more popular than the cars are the girls and the only easier ride than the girls are the cars'. In this programme, we find wet T-shirt contests, sisters 'who do everything together', 'hard girls' with 'fantastic jugs on display', and simulated sex on the dance floor ('she's dirty dancing', whilst her partner has just 'come dancing'). These women perform to crowds of baying, cheering men who reach out to touch the gyrating female forms: the women are dehumanized and commodified as they act out a frenzy of masculine fantasies.

The programme then shifts from the raucousness of Daytona Beach to Miami Beach, the American Riviera where apparently 'there's more girl-on-girl action ... than in a locker room in Wimbledon' and the camera focuses on several close up shots of women kissing each other. The narrator informs us that tourists come here for the weather, the light and the sensuality and 'because Miami girls are easier than a one piece jig saw puzzle'. In case we are slow to get the message, the next scene shows women dancing in a bar, where they 'really show their class; sorry not class, ass'. Moving the focus from the USA to Europe, this *Bikini Babes* programme's next stop is Sardinia, Italy, where the narrator says, the women are much like sardines: '... they're crammed in together, they're very tasty and they're often found hanging off a fisherman's rod'. As the camera pans over beaches peopled by topless sunbathing women and male onlookers, he continues 'very beautiful beaches attract very beautiful people'. Apparently, 'you can have every kind of fun in Sardinia' and as a bronzed, blonde topless woman walks past the camera, the narrator tells us that the island 'has a long oral tradition ... there's cunning linguists everywhere ... there's nothing a Sardinian girl loves more than a great big stud in her mouth (as the camera closes in on her pierced tongue).

For its final segment, the programme moves to one of the Hedonism resorts in the Caribbean. Hedonism II and III are adult-only vacation resorts in Jamaica operated by the company Superclubs (there is no Hedonism I). Hedonism II was opened in 1976 in Negril and Hedonism III followed in 1999 in Ocho Rios; both have areas reserved for naturism and nudism (although nudity is illegal in Jamaica these laws are not applied in the private resorts) and are well known for the liberal attitudes of their guests towards alcohol, drugs and sexuality. Here 'there are no rules ... the only rule is to drop your inhibitions'. To this end, pursuits such as naked drinking games, naked twister and nude bathing are the norm. Unlike the earlier segments of the programme, however, we see both men and women naked; apparently, 'Hedonism is great because it attracts all sorts ... you never normally see people like this wandering around naked'. The narrator says this just as the camera focuses on a naked woman in her forties with slightly sagging breasts and skin that has seen too much exposure to the sun. The narrator continues: 'maybe because you'd have closed your eyes and run screaming for cover'. In another programme in the series, the narrator comments '... if we know one thing about Hedonism ... it may be very grown up but it's not completely civilized,' voiced over footage of a very corpulent naked woman. In Hedonism, however, '... everyone is completely happy with their bodies ... no one cares who's got the best tits ... nobody's forced to do anything they don't want to do'. Whilst the tourists might be happy with their own bodies, the programme makes it obvious that these people do not conform to the perfect image of a 'hard earned beach body' and uses humour and ridicule to reinforce these bodily ideals. One middle-aged woman talks to the camera, saying 'you don't have to go naked if you don't want to' but her comments are interrupted by the narrator who says: 'look at her standing there, naked for all the world to see', implying that her body, whilst slender, is unattractive because of her age. By comparison, very little comment is made about the middle-aged naked men around her – as men they are the unremarkable norm and their bodies immune from any extended social criticism.

Conclusion

This chapter has explored how programmes such as *Wild On!* and *Naked Wild On! 2* celebrate 'low' cultural experiences in liminal zones which promise 'illicit or disdained social activities' (Shields 1991, 3). These tourism sites have received relatively scant attention in the heritage tourism studies literature which is dominated by studies of resource management (Timothy and Boyd 2006), which tends to privilege the high cultural tourist. We have seen here how such beach sites are a contemporary form of carnival (Bahktin 1984), marked by unrestrained excess and embodied celebrations. As such, these tourism spaces embody our 'consumption of space, sun and sea and of spontaneous or induced eroticism' (Lefebvre 1991, 58) and the exaggerated exoticism, eroticism, escapism and excess featured promote a form of sex tourism in these marginal, liminal spaces. This is particularly pertinent because, if material and imagined space is the location for tourism and leisure encounters, it is through our bodies that we experience, know and make sense of such places and personal interactions. The body is also the ultimate site of control and resistance and it has emerged in twenty-first century consumer culture as *the* significant symbolisation of self. It is a key metaphor for identity and the media, advertising and fashion industries, amongst others – including leisure and tourism – all contribute to the discourses within which we manage and locate our own bodies (Evans and Lee 2002). The body symbolizes the self, it connects us with other people and places but also marks us as different and 'out of place' (Cresswell 1996). The corporeal is therefore the pre-eminent site where our personal identities are constituted and social knowledges and meanings inscribed, and there are many ways in which bodies may be 'sexualized'. Moreover, liminal zones such as tourist resorts and night-time leisure spaces are closely identified with pleasure, desire, sex and sexuality. It is in these places that we see people actively engaged in (re) creating their sexual selves and performing their sexual identities (Butler 1990), which are themselves often entwined with contested social spaces. Certainly, spaces of pleasure and leisure create possibilities of alternative identities of self through what Fullager (2002, 65) has elsewhere characterized as the 'movement of release, of letting go in order to be open to difference through an experience of desire'. Desire is a deeply embedded, affective and unconscious imperative, which 'mediates our travel relations and experiences in culturally specific ways' (Fullager 2002, 57), yet in tourism studies, conceptualizations of desire have focused on individual motivation as opposed to a motivation to move 'towards that which is different, unknown and other to the self' (Fullager 2002, 57).

The beach and carnivalesque celebrations such as spring break provide performative stages of drama and role-playing, places where the meanings and the fluidities of our personal identities – themselves a series of multiple, multi-faceted reflexive performances – can be confirmed or (re)constituted. These beaches are liminal places – '... intangible, elusive, and obscure. They lie in a limbo-like space often beyond normal social and cultural constraints. In these spaces can be found brief moments of freedom and an escape from the daily grind of social responsibilities' (Preston-Whyte 2004, 350). Yet, although spaces like the beach and events like spring break offer a temporary release from everyday moral codes, they simultaneously reinforce the discipline of the body and gendered practices, so that people engage in 'deviant' practices 'safe in the knowledge that they are not transgressing the wider social structures they encounter in everyday life' (Ravenscroft and Gilchrist 2009, 36). And perhaps this is confirmed in that, as we have seen here, media representations of these spaces and festivals offer highly circumscribed constructions of femininity and masculinity. Women '... are explicitly positioned as the site of spectacle, display and consumption' (van Eeden 2007, 201), they are objectified as ornamental trophies, conforming to particular desirable bodily norms and conventions. At the same time, representations of masculinities offer opportunities for the 'acceleration of the ... signification of masculinity' (Know and Hannam 2007, 265). Beer drinking, leering over women, and celebrations of physicality through body sliding, belly flopping and sexual conquest offer distinct markers of masculinity, with little room for alternatives. Significantly, despite the increasing commodification of the male body in popular cultural forms, here male bodies are immune to scrutiny and are rarely remarked upon, in direct contrast to their female counterparts.

We began this chapter by reviewing how as individuals we are now able to exert more and more control over our bodies but how these bodies (particularly women's) are also much more vulnerable to control by others. Our media is saturated with body stories; thin, sexy and beautiful bodies are feted, whilst overweight bodies are chastised. Our bodies may now be objects for display but sexualisation and objectification undermine confidence in and comfort with one's own body, leading to a host of negative emotional consequences, such as shame, anxiety, and even self-disgust. Frequent exposure to media images that sexualize girls and women affects how girls conceptualize femininity and sexuality. These are the models of femininity which are presented to adults and children in today's world. Images of female bodies are everywhere. Women – and their body parts – sell everything from food to cars. Popular film and television actresses are becoming younger, taller and thinner. The barrage of messages about thinness, dieting and beauty tells 'ordinary' women that they are always in need of adjustment – and that the female body is an object to be perfected. As Susie Orbach (2009) has commented about the globalization of beauty, this has led to Iranian girls wanting Nicole Kidman's nose, American girls trying to be Kiera Knightley's dress size, Chinese women inserting rods in their legs to become taller and Brazilian women padding their bums to look like Jennifer Lopez.

Addressing the issue of the gendered and sexist visual representations that create these anxieties is, of course, not tourism-specific; rather the tourism sector is yet another example where traditional stereotypes come into play. So, in travel entertainment programmes such as those analysed here, the globalised tourism industry acts a key agent of cultural pedagogy, forming an element in the circle of representation (Hall 1997) we discussed in the introduction. In the same way as we need to develop more such examinations of travel's gendered visual rhetoric, we need to examine the cultural meanings and practices which surround embodied spaces, bodies in spaces, represented bodies, and the tourism consumer as a liminal, metamorphic body. The time is long overdue for a theorization of the cultural discourse of tourism spaces that engages with the visual and the linguistic, and with identity and the body. Tourism work elsewhere (see Pritchard et al. 2007) is construing the body as the ultimate site of social and cultural identity and yet heritage tourism studies (and tourism studies in general) have too rarely engaged critically with questions of gender, sexuality and the body. Even though the female body and its visual sexualization have been the focus of this chapter, this call for more focus on embodiment in tourism research (both in the material and symbolic domains), also asks for equal emphasis to be placed on male as well as female bodies. Ironically, much contemporary feminist research on the body has largely ignored the male body, thus paradoxically confirming 'the dualism which links bodies and bodily matters to women and femininity' (Davis 1997, 19).

References

Abramovici, M. (2007), 'The Sensual Embodiment of Italian Women', in A. Pritchard, N. Morgan, I. Ateljevic and C. Harris (eds) *Tourism and Gender: Embodiment, Sensuality and Experience*, 107–25. (Oxford: CABI).

American Psychology Association (2007), *Task Force on the Sexualisation of Girls*, [Online]. Available at: http://www.apa.org/pi/wpo/sexualization.html [accessed: 1 February 2008].

Attwood, F. (2006), 'Sexed Up: Theorizing the Sexualization of Culture', *Sexualities* 9:1, 77–94.

Bahktin, M. (1984), *Rabelais and His World.* (Bloomington: Indiana University Press).

Boyd, S.W. and Butler, R.W. (2000), 'Tourism and National Parks: The Origin of the Concept', in R.W. Butler and S.W. Boyd (eds) *Tourism and National Parks: Issues and Implications*, 13–27. (Chichester: Wiley).

Butler, J. (1990), *Gender Trouble: Feminism and the Subversion of Identity.* (London: Routledge).

Chernin, K. (1983), *Womansize: Tyranny of Slenderness.* (London: The Women's Press Ltd).

Cresswell, T. (1996), *In Place/Out of Place: Geography, Ideology and Transgression.* (Minneapolis, Minnesota: University of Minnesota Press).

Davis, K. (1997), 'Embody-ing Theory Beyond Modernist and Postmodernist Readings of the Body', in K. Davis (ed.) *Embodied Practices. Feminist Perspectives on the Body*, 1–23. (London: Sage Publications).

Esfahani, E. (2006), 'The New Skin Trade' [Online]. Available at: http://money.cnn.com/magazines/business2/business2_archive/2006/01/01/8368124/index.htm [accessed: 18 June 2009].

Evans, M. and Lee, E. (eds) (2002), *Real Bodies: A Sociological Introduction.* (New York: Palgrave).

Featherstone, M. (1991), 'The Mask of Ageing and the Postmodern Life Course', in M. Featherstone, M. Hepworth and B.S. Turner (eds) *The Body: Social Process and Cultural Theory*, 371–89. (London: Sage Publications).

Foucault, M. and Gordon, C. (ed.) (1982), *Power/Knowledge: Selected Interviews and Other Writings*. (New York: Pantheon).

Fredrickson, B.L. and Roberts, T-A. (1997), 'Objectification Theory: Toward Understanding Women's Lived Experience and Mental Health Risks', *Psychology of Women Quarterly* 21, 173–206.

Frosh, S., Phoenix, A. and Pattman, R. (2003), 'The Trouble with Boys', *The Psychologist* 16:2, 84–7.

Fullagar, S. (2002), 'Narratives of Travel: Desire and the Movement of Feminine Subjectivity', *Leisure Studies* 21:1, 57–74.

Gow, J. (1996), 'Reconsidering Gender Roles on MTV: Depictions in the Most Popular Music Videos of the Early 1990s', *Communication Reports* 9, 151–61.

Grauerholz, E. and King, A. (1997), 'Primetime Sexual Harassment', *Violence Against Women* 3, 129–48.

Gullelte, M.M. (1997), *Declining to Decline: Cultural Combat and the Politics of Midlife*. (Charlottesville and London: University Press of Virginia).

Hall, S. (1997), 'The Spectacle of the "Other"', in S. Hall (ed.) *Representation: Cultural Representation and Signifying Practice*, 223–90. (London: Sage Publications and the Open University).

Jackson, P., Stevenson, N. and Brooks, K. (2001), *Making Sense of Men's Magazines*. (Cambridge: Polity Press).

Jokinen, E. and Veijola, S. (1994), 'The Body in Tourism', *Theory, Culture* and *Society* 11:3, 125–51.

Jordan, F. (2007), 'Life's a Beach and then we Diet: Discourses of Tourism and the "Beach Body" in UK Women's Lifestyle Magazines', in A. Pritchard, N. Morgan, I. Ateljevic and C. Harris (eds) *Tourism and Gender: Embodiment, Sensuality and Experience*, 92–106. (Oxford: CABI).

Knox, D. and Hannam, K. (2007), 'Embodying Everyday Masculinities in Heritage Tourism(s)', in A. Pritchard, N. Morgan, I. Ateljevic and C. Harris (eds) *Tourism and Gender: Embodiment, Sensuality and Experience*, 263–72. (Oxford: CABI).

Krassas, N.R., Blauwkamp, J.M. and Wesselink, P. (2001), 'Boxing Helena and Corseting Eunice: Sexual Rhetoric in *Cosmopolitan* and *Playboy* Magazines', *Sex Roles* 44, 751–71.

Krassas, N.R., Blauwkamp, J.M. and Wesselink, P. (2003), '"Master Your Johnson": Sexual Rhetoric in *Maxim* and *Stuff* Magazines', *Sexuality and Culture* 7, 98–119.

Langman, L. (2003), 'Culture, Identity and Hegemony: The Body in the Global Age', *Current Sociology* 51, 223–47.

Lefebvre, H. (1991), *The Production of Space*, trans. N. Donaldson-Smith. (Oxford: Blackwell) (original edition 1974).

Lin, C. (1997), 'Beefcake Versus Cheesecake in the 1990s: Sexist Portrayals of Both Genders in Television Commercials', *Howard Journal of Communications* 8, 237–49.

Mackie, V. (2000), 'The Metropolitan Gaze: Travellers, Bodies and Spaces', *Intersections: Gender, History and Culture in the Asian Context*, 4. [Online]. Available at: http://intersections.anu.edu.au/issue4/vera.html [accessed: 2 August 2004].

McKinley, N.M. and Hyde, J.S. (1996), 'The Objectified Body Consciousness Scale', *Psychology of Women Quarterly* 20, 181–215.

Morgan, N. and Pritchard, A. (1998), *Tourism Promotion and Power: Creating Images, Creating Identities*. (Chichester: Wiley).

Orbach, S. (2009), *Bodies*. (London: Profile).

Parker, R. (1991), *Bodies, Pleasures and Passions: Sexual Culture in Contemporary Brazil*. (Boston: Beacon Press).

Preston-Whyte, R. (2004), 'The Beach as a Liminal Space', in A. Lew, C.M. Hall and A. Williams (eds) *The Blackwell's Tourism Companion*, 249–59. (Oxford: Blackwells).

Pritchard, A. (2001), 'Tourism and Representation: A Scale for Measuring Gendered Portrayals', *Leisure Studies* 20:3, 79–94.

Pritchard, A. and Morgan, N. (2000a), 'Privileging the Male Gaze: Gendered Tourism Landscapes', *Annals of Tourism Research* 27:3, 884–905.

Pritchard, A. and Morgan, N. (2000b), 'Constructing Tourism Landscapes: Gender, Sexuality and Space', *Tourism Geographies* 2:2, 115–39.

Pritchard, A. and Morgan, N. (2005), 'On Location: (Re)Viewing Bodies of Fashion and Places of Desire', *Tourist Studies* 5:3, 283–302.

Pritchard, A. and Morgan, N. (2007), 'Encountering Scopophillia, Sensuality and Desire: Engendering Tahiti in Travel Magazines', in A. Pritchard, N. Morgan, I. Ateljevic and C. Harris (eds) *Tourism and Gender: Embodiment, Sensuality and Experience*, 158–81. (Oxford: CABI).

Pritchard, A., Morgan, N., Ateljevic, I. and Harris, C. (eds) (2007), *Tourism and Gender: Embodiment, Sensuality and Experience.* (Oxford: CABI).

Quine, S., Bernard, D. and Booth, M. (2003), 'Locational Variation in Adolescents' Concerns over "Body Image"', *Health Promotion Journal of Australia* 14:3, 224–5.

Ravenscroft, A. and Gilchrist, P. (2009), 'Spaces of Transgression: Governance, Discipline and Reworking the Carnivalesque', *Leisure Studies* 28:1, 35–50.

Rojek, C. and Urry, J. (eds) (1997) *Touring Cultures: Transformations of Travel and Theory.* (London: Routledge).

Ryan, C. and Hall, C.M. (2001), *Sex Tourism: Marginal People and Liminalities*. (London: Routledge).

Savage, J. (1990), 'Tainted Love: The Influence of Male homosexuality and Sexual Divergence on Pop Music and Culture Since the War', in A. Tomlinson (ed.) *Consumption, Identity and Style*, 153–71. (London: Routledge).

Shields, R. (1991), *Places on the Margin: Alternative Geographies of Modernity*. (London: Routledge).

Shilling, C. (2003), *The Body and Social Theory*. (London: Sage).

Slater, A. and Tiggemann, M. (2002), 'A Test of Objectification Theory in Adolescent Girls', *Sex Roles* 46, 343–9.

Small, J. (2007), 'The Emergence of the Body in the Holiday Accounts of Women and Girls', in A. Pritchard, N. Morgan, I. Ateljevic and C. Harris (eds) *Tourism and Gender: Embodiment, Sensuality and Experience*, 73–91. (Oxford: CABI).

Timothy, D.J. and Boyd, S.W. (2006), 'Heritage Tourism in the 21st Century: Valued Traditions and New Perspectives', *Journal of Heritage Tourism* 1:1, 1–16.

Tomlinson, A. (1991), (ed.) *Consumption, Identity and Style*. (London: Routledge).

Turner, V. (1974), *Dramas, Fields and Metaphors*. (Ithaca, NY: Cornell University Press).

Urry, J. (2002), *The Tourist Gaze*. (London: Sage Publications).

van Eeden, J. (2007), 'Gendered Tourism Space: A South African perspective', in A. Pritchard, N.Morgan, I. Ateljevic and C. Harris (eds) *Tourism and Gender: Embodiment, Sensuality and Experience*, 182–206 (Oxford: CABI).

van Gennep, A. (1960), *The Rites of Passage*, trans. M.B. Vizedom and G.L. Caffe. (Chicago: University of Chicago Press).

Veijola, S. and Jokinen, E. (1994), 'The Body in Tourism', *Theory, Culture and Society* 11, 125–51.

Vincent, R.C. (1989), 'Clio's Consciousness Raised? Portrayal of Women in Rock Videos, Re-examined', *Journalism Quarterly* 66, 155–60.

Walton, J. (2000), *The British Seaside Resort*. (Manchester: University of Manchester Press).

Ward, L.M. (2004), 'Wading Through the Stereotypes: Positive and Negative Associations Between Media Use and Black Adolescents' Conceptions of Self', *Developmental Psychology* 40, 284–94.

Ward, L.M. and Rivadeneyra, R. (1999), 'Contributions of Entertainment Television to Adolescents' Sexual Attitudes and Expectations: The Role of Viewing Amount Versus Viewer Involvement', *Journal of Sex Research* 36, 237–49.

Westwood, S.H. (2004), 'Narratives of Tourism Experiences: An Interpretative Approach to Understanding Tourist-Brand Relationships', unpublished PhD thesis, University of Wales Institute, Cardiff.

Williamson, J. (1986), *Consuming Passions: The Dynamics of Popular Culture*. (New York: Marion Boyers).

Zurbriggen, E.L. and Morgan, E.M. (2006), 'Who Wants to Marry a Millionaire? Reality Dating Television Programs, Attitudes Toward Sex, and Sexual Behaviors' *Sex Roles* 54, 1–17.

Chapter 9
Authenticity, the Media and Heritage Tourism: Robin Hood and Brother Cadfael as Midlands Tourist Magnets

Roy Jones

Eco (1986) has pointed out that many tourists seek hyper-real destinations. One means by which such hyper-reality can be attained and/or heightened, particularly in visual terms, is through the link between mythical or fictional characters and their supposed actions in actual places, and especially in places with heritage significance. Two such places where this process can be observed are Shrewsbury Abbey in Shropshire and Sherwood Forest in Nottinghamshire, both in the English Midlands. Shrewsbury Abbey is the fictional home of Brother Cadfael, a twelfth-century monk-detective and the hero of numerous novels and television programs. Sherwood Forest was first associated with the mythical actions of Robin Hood in late medieval ballads, and the media, in many forms, has continuously built upon this legend over the intervening centuries. At both locations, these fictional/mythical characters have been used to develop local tourism industries and even to rehabilitate and visually transform areas suffering from economic and/or environmental problems. Using Selwyn's (1996) concepts of 'hot' and 'cool' authenticity in tourism, this chapter will summarize both the histories and the fictional/mythical overtones of these areas before considering how these have been blended in the construction of local tourist narratives and the visual culture that supports these.

As demonstrated by the popularity of 'battlefield tourism' (McGreevy 1991) at many locations from Gallipoli to Gettysburg, or of 'dark tourism' (Lennon and Foley 2000) more generally, tourism sites can benefit considerably from an association with violent (or romantic) events and heroic (or notorious) characters – sites, in short, where both the actual landscapes and the events associated with them conjure up vivid pictures in the eye and mind. Furthermore, it would seem that these tourism benefits can often accrue to places, even when the events and the characters with which they are associated are mythical and/or fictional. The presence of Mount Olympus is hardly a drawback for tourists to Greece, though very few now see it as the home of the Gods; likewise, Prince Edward Island's tourist industry depends very heavily on the Anne of Green Gables stories. Modern media events can also play a significant, though not necessarily deliberate, role in tourism as aficionados seek out the locations that they have seen in successful films and television series (see Mordue 1999 on 'Heartbeat Country'). For example, although it is, in any case, a tourist 'historic gem' of a town (Ashworth and Tunbridge 2000), Stamford benefited considerably from being the location for the BBC's 1994 televisation of *Middlemarch*, as did Lyme Hall, the stately home at which 'Mr Darcy' (Colin Firth) famously emerged from a lake in the BBC's adaptation of *Pride and Prejudice.*

Selwyn, in his work on tourism and myths, considered the power of history and fiction, and distinguished between the representation of 'cool' and 'hot' authenticity, both visually and in more general terms, as follows:

> The suggestion here is to work with two distinct senses of the term 'authentic'. A first type may be termed 'hot authenticity' and will apply to that aspect of the imagined world of tourist make-believe – that aspect of tourist myths – concerned with questions of self and society. The unashamedly modernist suggestion is that underneath the surface structures of the post-modern tourist myths … are modern and even pre-modern concerns with the 'authentic self' and the 'authentic other' … A second type 'cool authenticity' may be reserved for propositions which aim to be open to the kinds of procedures described by Popper … and which would like to claim a different kind of legitimacy from those in the former category (1996, 20–1).

Cool authenticity, therefore, refers to that which can be empirically demonstrated to have occurred, to buildings and/or sites, still extant and visible, of proven antiquity and, ideally, to sites where people, who are generally considered to be significant contributors to national, or at least local, heritages, incontrovertibly performed notable feats. Hot authenticity, by contrast, produces responses from the emotions, rather than from the intellect, responses, perhaps, to issues of good and evil, or even just to a good story. For example, tours focusing on ghosts and murders are becoming increasingly common across the Western world. But tourism and, even more so, place marketing relate to authenticity in a more complex manner than that indicated by the simple dichotomy between the factual (cool) and the fabulous (hot). Even where actual and colourful characters were involved in real, spectacular and nationally significant events (such as Napoleon and Wellington at Waterloo), a whole range of media representations, from *Sharpe's Waterloo* through *Vanity Fair* to historical documentaries, will increase public awareness and interest, and thus stimulate tourist movements.

At a further level of complexity, fictional or mythical stories are often set in actual locations. The Colosseum in Rome (already, of course, a major tourist attraction) featured strongly in the highly successful film *Gladiator*, which was a classic piece of hot authenticity in terms of its dramatic depictions of good and evil. My frequent use of (real and fictional) examples from both the distant and the not so distant past is deliberate. Selwyn is not alone in pointing out that, insofar as people seek for their 'authentic' selves through tourism, they often focus on those earlier (or different) societies, which they perceive as offering them simpler and, thus, more 'real' experiences than do their own modern, urban environments (see also Cohen 1994; MacCannell 1999, 91). It matters not that their Indigenous, farm stay, 'pioneer' or other historical tourist experiences are frequently restored and sanitized to the point of what Eco (1986) terms hyper-reality or that the more aesthetic aspects of the visual are privileged over the less pleasant aspects of the earlier aural and olfactory realities. Indeed, the very unreality of what is being perceived may assist the viewers/tourists in relating the contemporary visual experience with their own cultural or national heritages

English history and (again according to Eco) medieval history are, perhaps, particularly evocative both in terms of hot authenticity, through what we perceive as their heady mix of chivalry and violence and, in terms of cold authenticity, through the visual impressiveness of their physical remains (cathedrals, town walls, castles, abbeys etc.). In this chapter, I will, therefore, focus on two areas in the English Midlands, Shrewsbury in Shropshire and Sherwood Forest in Nottinghamshire, where fictional/mythical/media-generated medieval figures have been co-opted, not merely into becoming the bases of local heritage tourism attractions, but also into being catalysts for place marketing and even for local economic regeneration. In both cases, I will summarize firstly the cool authenticity of local historical record, and secondly the hot authenticity of the exploits of, the fictional monk-detective, Brother Cadfael of Shrewsbury Abbey, and the mythical outlaw, Robin Hood of Sherwood, respectively. I will then consider the more complex (tepid?) authenticity of the development of local tourist industries which have both used and co-opted a variety of essentially visual media to blend these cool/real (Shrewsbury Abbey and Sherwood Forest) and hot/unreal (Brother Cadfael and Robin Hood) components into heritage tourism attractions.

Cool Authenticity

Shrewsbury Abbey/Abbey Foregate

As a frontier fortress town close to the border with Wales, Shrewsbury, in cold historical fact, experienced a turbulent and frequently violent history throughout the Middle Ages, a fact that was first acknowledged in 'hot authenticity' terms when Shakespeare made the Battle of Shrewsbury in 1403 the climax of *Henry IV, Part 1*.

At the beginning of this period, in 1083, The Abbey of St Peter and St Paul in Shrewsbury was founded by French, Benedictine monks and was probably built on the site of an earlier wooden church. Like most abbey communities at that time, the monks were the caretakers of holy relics, in this case the remains of St Winifred, which attracted pilgrims to the abbey and the town. At that time, Shrewsbury was a major fortress and county town guarding the borders of England from attacks by the Welsh. A meander of the River Severn offered Shrewsbury its protection, within which the town and its walls were built, with a castle guarding the meander's narrow neck. The Abbey was constructed immediately East of the town next to the gated English Bridge. The major road from London (which roughly followed the route of the Roman road known as Watling Street) passed by the Abbey and entered the town over the bridge. Although much of the land near the Abbey was taken up by this community's own gardens, ponds and a mill on an adjacent stream, the suburb of Abbey Foregate began to develop around it in the early Middle Ages.

Given this location, the Abbey would have certainly been a focal point in the movements of many people in this frontier place in these uncertain times. The protection offered by the town and its walls was frequently required during the Middle Ages when Shrewsbury was besieged several times. A notable siege occurred during the mid-twelfth century, the period during which the first Brother Cadfael novels are set. England was then experiencing a civil war between King Stephen and the Empress Maud/Matilda who both claimed the crown.

Since the Middle Ages, both the abbey itself and the surrounding suburb of Abbey Foregate have experienced considerable change. During the dissolution of the monasteries in the sixteenth century, the monastic buildings at Shrewsbury were destroyed, though the Abbey Church itself remained. Charles II offered to make Shrewsbury a city and the Abbey Church a cathedral in thanks for the town's support during the English Civil War. However, his offer was rejected by the civic leaders of the time on the grounds that such an elevation had not been the reason for their support of the Royalists.

Thomas Telford, the pioneer civil engineer, constructed the London-Holyhead ('Irish Mail') turnpike road through the former monastic site in the eighteenth century. In the nineteenth century, Shrewsbury also became a major railway junction, and a railway station was subsequently built on the site of the monastic buildings. Abbey Foregate was crossed by railway lines and a large locomotive depot was constructed close by. Even the Abbey's former gardens (the Gaye) on the banks of the Severn became the Gay Meadow, the stadium of the local football club up to the 2006–7 season. By the late twentieth century, Abbey Foregate railway station had been closed, road congestion approaching the English Bridge was severe and the area surrounding the Abbey was suffering from urban blight.

Sherwood Forest/Edwinstowe

Edwinstowe allegedly (this word recurs constantly with reference to the village and its surrounds) developed around the burial site of Edwin, King of Northumbria, who was killed in battle against the Mercians in the early seventh century and buried in the forest at the battle site. Subsequently, he was sanctified and a chapel was constructed at the site of his tomb. By 1086, a small village was grouped around this religious building ('a church, a priest and four bordars' according to the Domesday Book) and the first stone church was built there in 1175.

Edwinstowe was formerly within, and is currently adjacent to, (a much reduced) Sherwood Forest. Sherwood Forest occupies a tract of land in the North of Nottinghamshire, where the soil was too poor to merit its early clearance for agriculture. It was not necessarily thickly wooded, and parts of it would have been sandy heathland. In the early Middle Ages it was therefore classed as common land where local people could hunt, gather, collect fuel and graze their stock. When the Norman monarchs designated this area a royal forest, these public privileges were withdrawn – coinciding with restrictions placed upon public access to crown land throughout the country at that time – a move which caused popular resentment locally and nationally. However, the resentment was, perhaps, personalized to a greater extent in the area around Edwinstowe when one of the 'coolly' authentic characters of the Robin Hood legend, King John, built a hunting lodge there.

During and after the late Middle Ages, as the population grew and as farming techniques improved, the forested area was gradually reduced. However, it is perhaps a reflection of the relatively poor agricultural quality of the land that, in the seventeenth and eighteenth centuries, the area North and East of the remaining forest became 'The Dukeries', a region of stately homes and extensive landscaped parks. The region then experienced extensive environmental and socioeconomic disruption in the late nineteenth and early twentieth centuries as the Nottinghamshire coalfield was developed. Pre-existing villages, including 'Robin Hood's village' of Edwinstowe, were massively expanded to accommodate the miners and their families. By the turn of the millennium, however, most of the pits had closed and the area was experiencing high unemployment and levels of socioeconomic distress.

Hot Authenticity

Brother Cadfael

The fictional character Brother Cadfael was a Benedictine monk at Shrewsbury Abbey around 1140 and was the detective hero of 20 novels (termed 'medieval whodunits' by their publisher) and three short stories. These murder mysteries were published, roughly annually, from 1977 to 1995 – the year of the death of the author, Edith Pargeter, a local woman who used the pseudonym Ellis Peters when writing her detective stories. Between 1994 and 1998, four series of television dramas based on the novels were made, largely on location in Hungary. The profile of the series was enhanced by the presence of the noted Shakespearian actor, Sir Derek Jacobi in the role of Brother Cadfael. These programs were shown internationally and appeared on the Public Broadcasting Service in the USA and on a commercial channel in Australia.

While Shrewsbury and the Abbey were the prime foci of these works, their 'action' frequently extended across Shropshire and its adjacent counties. Characteristically, the books included maps to enable readers to follow their plots. Notwithstanding the antiquity of their settings, it is still possible for anyone visiting or with a knowledge of these areas to pinpoint, and to a certain extent, therefore, to visualize an impressively large proportion of the locations used in these stories.

Robin Hood

The Robin Hood stories are set in the final years of the twelfth century (Dobson and Taylor 1972). According to the Nottinghamshire County Council website, in 1261, an outlaw in Buckinghamshire was given the name of 'William Robinhood' by an official recording his case. Eight such cases of nicknaming had occurred by 1300. Schama (1995) has also noted that one Robert Hood of Wakefield, Yorkshire, was convicted of taking wood from the earl's forest in 1308 and was obliged to make a payment for it.

The earliest surviving poem about Robin Hood dates from 1400 and is preserved in Lincoln Cathedral. In the late fifteenth century, the 'Lytell Geste of Robin Hode' first appeared, which was a collection of songs about Robin Hood and his associates. Apparently, these had first been performed as court ballads, but, through the then dominant oral tradition, the range of songs became more extensive and the range of their performers and listeners gradually became more socially inclusive. Initially, the legend of these 'gestes' was pro-establishment. Robin Hood was portrayed as standing up for the king and for the people's ancient rights against the corrupt officials who sought to profit from Richard I's absence during the crusades. Certainly Henry VIII took part in Robin Hood festivities in 1515, complete with a venison breakfast out of doors and 200 Lincoln green-clad archers. Later that century, Shakespeare compared the situation of his characters in the Forest of Arden (also in the English Midlands) in *As You Like It* with that of Robin Hood and his band.

However, in 1795, Joseph Ritson, who was writing at the time of, and was in sympathy with, many of the ideals of the French Revolution, published the two volume *Robin Hood: A Collection of All the Ancient Songs and Ballads*, in which a more radical dimension of 'robbing the rich to help the poor' was added to the Robin Hood legends. Since Sir Walter Scott consulted both Ritson the man and Ritson's works while writing his medieval novels, this political view of the Robin Hood legend became more widely popularized, and some twentieth-century treatments of the story took an almost communist approach to the adventures of Robin and his 'comrades'.

For much of the last century, a steady stream of film and television treatments of the legend has complemented the print and dramatic material, with Robin Hood being represented on screen by a wide range of international actors from the Australian Errol Flynn to the American Kevin Costner. In recent years, the legend has been made increasingly inclusive in other dimensions, with, for example, a black outlaw featuring in the film *Robin Hood: Prince of Thieves* and a Muslim, female outlaw in the recent BBC TV production, not to mention such comedic treatments as *Robin Hood: Men in Tights* and even *Maid Marian and her Merry Men.* Each of these representations draws on a conventional series of visual cues, such as the green costumes, the forest setting, and the primacy of the longbow as the weapon of choice. Although Yorkshire figures in several of the stories, including that of his death, and that county contains a Robin Hood's Bay, most of the legends focus on Sherwood Forest and on Nottinghamshire. Notwithstanding the centrality of Sherwood/Nottinghamshire to the Robin Hood legends, it must also be acknowledged that they are now also a significant part of a broader English heritage in which the (virtuous, English) Saxons are opposed to the (corrupt, French) Normans. Even more broadly – and fortuitously in tourism terms – Robin Hood is also a classic example of the bandit hero (Hobsbawm 2001). Such figures (for example, Jesse James, Ned Kelly and Zorro) figure strongly in the cultural heritages of many nations and place Robin Hood in a near-universal genre in much the same way as Brother Cadfael fits into the popular category of detective fiction.

Tepid Authenticity

The Shrewsbury Quest and Cadfael Country

Shrewsbury has been a noted tourist destination virtually since it developed railway connections in the nineteenth century. The initial focus of this industry was on the historic town centre, with its exceptional collection of 'black and white' Tudor buildings and on the landscaped riverside parks and gardens, which are the site of one of England's most famous flower shows. By the 1980s, however, the 'Brother Cadfael' books were being sold at the Abbey's book and souvenir shop and, by 1990, a booklet of walks in Abbey Foregate and central Shrewsbury (printed by 'Brother Cadfael Products Limited') traced 'the footsteps of Brother Cadfael' on his various murder investigations. Soon after, a series of metal footprints were placed in the town's pavements to mark these 'footsteps'. Major tourism development in Abbey Foregate did not begin until the early 1990s, when land opposite the Abbey became available as part of the redevelopment of the former station yard. The Shrewsbury Quest (with its slogan 'Live the History … Solve the Mystery'), which was constructed on this site, contained a recreation of an abbey scriptorium, guest hall and herb garden (complete with the hut where Brother Cadfael prepared his medicines). Visitors could learn about elements of monastic and medieval life and, at the same time, find clues to solve a murder. They could also stand in the reconstructed buildings and gardens of the Shrewsbury Quest and see the medieval tower and aisle of the Abbey Church (but not the road that lay between the two; the 'hot' visual had primacy over cold authenticity). The centre was opened in 1994 with the support and advice of Edith Pargeter, and included a recreation of her study. In 1997, a memorial window to Edith Pargeter was unveiled in the Abbey church, perhaps fittingly in view of the major contributions made to its upkeep by Brother Cadfael tourists in recent years.

The Shrewsbury Quest closed in 2001. This was, perhaps significantly, a few years after the supply of new Brother Cadfael books and television programmes ceased. However, a member of the local Chamber of Commerce informed me that visitor numbers were consistently high and that other financial problems brought about its closure. In spite of the demise of the Shrewsbury Quest, the tourism impact of Brother Cadfael endures and extends well beyond Abbey Foregate. The Abbey still sells the Cadfael books and other memorabilia (which now include Cadfael herbals and recipe books and guides to the places mentioned in the novels). Interestingly, the Abbey is now addressing authenticity issues of its own. Signage clearly identifies the medieval/monastic (and thus Cadfael-related) parts of the building and distinguishes these from the (largely Victorian) repaired and reconstituted sections. Both the monk and his alter ego, Sir Derek Jacobi, have been used extensively to promote the tourist attractions of Shrewsbury, Shropshire and the entire region. In a tourism brochure for Shropshire, Jacobi (photographed tonsured and wearing a habit) is quoted as giving his 'blessing' to the Brother Cadfael Trail in Shrewsbury. Guide books and postcards now refer to Shropshire as 'Cadfael Country', in part displacing the county's other literary tourism magnet, A.E. Housman, the author of the popular 'Shropshire Lad' poems. The visual medium of television is now perhaps a more powerful attractant than the verbal medium of poetry.

Sherwood Forest and Robin Hood Country

In the late eighteenth century, while he was assessing Sherwood's timber resources for the Royal Navy during the Napoleonic Wars, Major Heyman Rook noted an oak so large and so old that it could have been there in Robin Hood's day. His comments caused the tree, which, with a certain lack of modesty, he named the Major Oak, to become romantically associated with Robin Hood. It was a secondary point that this tree would have been too small to provide either shelter or a hiding place for outlaws in the twelfth century. Following the so-called rediscovery of the Middle Ages in the nineteenth century, and the medievalism promoted by Ruskin, Morris and the pre-Raphaelites, the Major Oak began to be visited by the curious. Edwinstowe, at that time, was a pretty, rural village unaffected by coal mining, and it became a stopping off point for these visits. There were clearly local benefits in this development and the village and its church became increasingly complicit in the legend. The parish church had been new in Robin Hood's time and it therefore became the 'alleged' site of Robin Hood's marriage to Maid Marian. By the 1890s, Edwinstowe had its own railway access and the large Dukeries Hotel was established to cater for those tourists who wished to stay longer in the area.

The rise of the coal industry may have lessened the village's rural and historic charm, albeit temporarily, but its fall has certainly triggered a new interest in tourism as an alternative local economic base, with the Nottinghamshire County Council developing the Sherwood Forest Country Park and Visitor Centre on the edge of Edwinstowe. This is now a tourist site of national significance. The Major Oak (now kept alive by miracles of tree surgery) is its most notable attraction and its displays and exhibitions focus on forest ecology and on the Robin Hood legend, in both cases feeding off the, perhaps convenient, status of the oak as England's 'heritage tree'. Not only are many private gift shops, restaurants etc. in Edwinstowe (such as the 'Robin Hood's Plaice' fish and chip shop) heavily dependant on the Robin Hood tourist trade, but the county, district and even European governments have invested development funds in several tourism-related projects such as a craft centre and a Youth Hostel. In the adjoining village of Walesby, *The World of Robin Hood* exhibition added some props and costumes from *Robin Hood: Prince of Thieves* to a former Crusades exhibition brought in from Southern England, but, perhaps with the passing of Costner's film into history this, like the actor, has lost its drawing power and has recently closed.

As in Shropshire, the medieval 'brand' has been appropriated by the whole county, rather than by the immediate locality. Nottingham, of course, has the Sheriff's castle and its own mini theme park, *The Tales of Robin Hood*, which proudly states in its brochure that *The Independent* has called it 'a little Disneyland'. More generally, however, the County Library began to develop a Robin Hood archive as long ago as 1868. Nottinghamshire County Council brands itself as 'Robin Hood Country' and uses both the oak tree and the colour green widely in its 'corporate style'.

Conclusion

Both the examples discussed here clearly illustrate the symbiotic relationship between the tourism and the media industries. This has been widely noted in the literature and is evident from Avoca in Ireland (*Ballykissangel*) and Talkeetna, Alaska (*Northern Exposure*) to Barwon Heads in Australia (*Seachange*). Why this relationship has been particularly successful in the two instances documented here is, I would argue, a feature of time, at least as much as that of place. It may be more than coincidence that Eco (1986) juxtaposes his essays on *Travels in Hyperreality* and *The Return of the Middle Ages*. He considers the 'renewed interest in the Middle Ages to be a curious oscillation between fantastic medievalism and responsible philological examination' (1986, 63) – or perhaps between Selwyn's 'Hot' and 'Cool' authenticities. More specifically he identifies several 'little Middle Ages', which still attract our interest today, both as tourists and more generally. These include:

> Pretext – a mythological stage on which to place contemporary characters; Ironical Visitation – in the same way that Sergio Leone and the other masters of the 'spaghetti western' revisit nineteenth century America; A Barbaric Age – of elementary and outlaw feelings; Romanticism; National Identities – a celebration of past grandeur; Decadentism; Philological Reconstruction; So-called Tradition (Eco 1986, 63).

In short, Eco identifies the Middle Ages as both visually and dramatically engaging and as central to the development of national and cultural heritages. Clearly, romances and mysteries such as those of Robin Hood and Brother Cadfael, are particularly likely to appeal to our visual and emotional senses of all of these 'Little Middle Ages'. As such, they provide a very effective tourist 'hook', not only to attract visitors to these areas, but also to contribute to their economic, and even landscape, regeneration. Many of the former mine sites are being reforested and cloned Major Oaks are a popular item at the Sherwood Forest Country Park. With the construction of a bypass and the recent movement of the football ground, Abbey Foregate is now gentrifying. These tourist centres also act as foci for tourism promotion over wider areas covering much of Nottinghamshire and Shropshire. Clearly, in these cases of literary and media generated tourism, it is cool to be tepid.

References

Ashworth, G.J. and Tunbridge, J.E. (2000), *The Tourist-Historic City: Retrospect and Prospect of Managing the Heritage City.* (Oxford: Elsevier Science).

Cohen, E. (1994), 'Contemporary Tourism – Trends and Challenges: Sustained Authenticity or Contrived Post-modernity?', in R. Butler and D. Pearce (eds) *Change in Tourism*, 12–29. (London: Routledge).

Dobson, R.B. and Taylor, J. (1972), 'The Medieval Origins of the Robin Hood Legend: A Reassessment', *Northern History* 7, 1–30.

Eco, U. (1986), *Travels in Hyperreality: Essays.* (San Diego: Harcourt Brace Jovanovich).

Hobsbawm, E. (2001), *Bandits.* (London: Abacus).

Lennon, J. and Foley, M. (2000), *Dark Tourism.* (London: Continuum).

MacCannell, D. (1999), *The Tourist: A New Theory of the Leisure Class.* (Berkeley: University of California Press).

McGreevy, P. (1991), 'Review of A.J. Lamme III, America's Historic Landscapes: Community Power and the Preservation of Four National Historic Sites', *Environment and Planning A* 23:9, 1377–8.

Mordue, T. (1999), '"Heartbeat Country": Conflicting Values, Coinciding Visions', *Environment and Planning A* 31:4, 629–46.

Schama, S. (1995), *Landscape and Memory.* (London: Random House).

Selwyn, T. (1996), 'Introduction', in T. Selwyn (ed.) *The Tourist Image: Myths and Myth Making in Tourism*, 1–32. (Chichester: Wiley).

Chapter 10
Branding the Past: The Visual Imagery of England's Heritage

Emma Waterton

As John Urry (1994, xx) points out, '[i]dentity almost everywhere has to be produced partly out of the images constructed or reproduced for tourists'. It is therefore no surprise that a robust field of study concerned with 'the visual' has developed within the social sciences (Rose 2001, 10) and indeed within the study of tourism itself (Crouch and Lübbren 2003). This development, however, has had relatively little effect on studies of heritage, an area of research that encounters the visual constantly yet has generated very little in the way of literature dealing with issues of visual representation. This omission is surprising, especially as England's heritage and its meaning, as Stuart Hall (2005, 24 emphasis in original) reminds us, 'is constructed *within*, not above or outside representations', such that 'those who cannot see themselves reflected in its mirror cannot properly "belong"'. As an observation, Hall's statement is particularly acute when placed within the context of contemporary calls for social inclusion and multiculturalism. Indeed, it means that the 'visual' cannot simply be written off as a passing intellectual fad, for it allows us to address real issues in the political, social and cultural arenas.

In conforming to Hall's premise, I want to argue that the images used to promote heritage in England can be criticized for their abiding non-inclusiveness, and are likely to represent and construct a very particular idea of heritage that is both potent and self-fulfilling. This construction of heritage, the chapter will argue, speaks to – and is fundamentally *about* – the cultural symbols of an elite social group: the white, middle-classes. Nonetheless, this understanding of heritage is presented as a consensual past, such that heritage touristic places become ideological spaces within which experiences are presented as conflict-free, sanitized, focussed on leisure and predominantly family orientated. It is this idea of 'family spaces', as Durrheim and Dixon (2001, 441) note, that is used to privilege the experiences of one social class over others. As such, heritage tourism images become constructions that work specifically to 'exclude' as a consequence of their promotion of a distinctly white, middle-class experience. This trope of 'family space' is thus a hidden trace against which non-white, non-middle-class groups are subjected to a process of *othering*.

This chapter adopts in many ways, therefore, the sort of argument that is becoming more usual within heritage literature, with its focus on power and ideology. It is also, however, an attempt to strike out towards a more useful understanding of how the visual is apprehended, particularly its ability to both constitute and sustain powerful asymmetries within sociocultural structures. It is in many ways speculative, both in terms of its objectives and how it proceeds, but that is not to say that it is without a wider evidence base. Research by Morgan and Pritchard (1998), Palmer (1999, 2005), Pritchard and Morgan (2001), Jenkins (2003) and Buzinde et al. (2006), for example, have been synthesized here, and are supplemented with an examination of recent tourism representations and branding campaigns designed for England. Specific images used to capture the public's attention are explored, with emphasis placed upon *English Heritage Members' and Visitors' Handbooks* (editions 1999/2000, 2003/2004 and 2007/2008), the VisitBritain publication *Heritage Britain* (2004), and a number of custodianship brochures produced by English Heritage. All of these examples can be characterized by a desire to 'sell' a particular image of heritage. They therefore share a clear generic structure that revolves around marketing and commoditization, and represent what Wernick (1991, cited in Fairclough 2003, 112) views as 'promotional culture', allowing certain messages to co-exist, seamlessly and simultaneously, with 'our produced symbolic world'.

Reframing Heritage Imagery: Discourse and Critical Analysis

Using resources such as tourism brochures and websites as data for indexing and interrogating the nature of visual imagery, this chapter takes its cue from developments in the field of discourse analysis and the works of Norman Fairclough, Theo van Leeuwen and Gunther Kress. Each of these authors is directed by three important observations, all of which can be modified here so as to speak specifically of heritage: first, heritage is dynamically constructed and constituted in discourse (Benwell and Stokoe 2006, 4). Second, images of heritage, in the same way as texts, act as indicators of underlying conceptualizations, amid a meaning-making process that can both mask and obfuscate the affects of power, authority and control (Emmison and Smith, 2000, 58; Jenkins 2003). Third, and perhaps most importantly, these constitutive features of discourse and visual imagery play a key role in both sustaining and constructing that which they represent. In other words, visual imagery does more than provide a pictorial 'label' for heritage: it creates, promotes and preserves a particular vision of heritage as reality.

This framing is instructive of a social constructivist approach and, as I will inevitably draw upon notions of power and ideology, finds congruence with the methodological specificities of critical discourse analysis (henceforth CDA). In embryo, CDA is concerned with unpacking 'ideological discursive formations' (IDFs) and denaturalizing dominant IDFs whose underpinnings we might at first be unaware of (Fairclough 1995; Jaworski and Coupland 1999, 34). Any analysis adopting a CDA approach should, as Kress and van Leeuwen (2006, 14; Fairclough 2001b, 230) note, therefore be *critical*, and those attending to the 'visual' are no different: images are, after all, continually performing ideological roles. Thus, the apparently neutral and/or informative agendas of tourism brochures, guides and postcards are just as implicated in the processes of conveying power and status as more explicitly ideological texts (Kress and van Leeuwen 2006, 14). Moreover, they are, as Kress and van Leeuwen demonstrate, as much a topic of analysis and an avenue through which we might tackle wider social issues as verbal texts and language.

An in-depth introduction to the mechanisms and vocabularies central to CDA is not necessary here. Rather, it is more important to point to the analytical *direction* this underpinning might initiate, particularly in terms of how we might unpack the 'grammar of visual design' (Kress and van Leeuwen 2006, 15). The point to draw out of the associated literature, then, is that the (re)materialization of particular discourses affects their operationalisation, in terms of subsequent ways of acting and interacting (genres), inculcation, or 'ways of being' (styles), and the representations (discourses) it privileges (Fairclough 2001a, 29). Each act of communication, whether visual or not, will thus always do something to, for or with others in the present (Kress and van Leeuwen 2006). Using discourse analytical techniques of the visual therefore means placing attention upon elements of an image such as the social distance implied between viewer and image, the level of interaction or involvement suggested, the camera angles/height used and the generalization/specificity of the images (Waterton 2009).

Given the ubiquity of the visual within the realm of heritage studies, this analysis takes interest in a particular visual narrative, or IDF, and my goal is to examine a succession of instances within which this IDF reveals its capacity to naturalize and gain acceptance for a particular conceptualization of heritage (Fairclough 1995, 27). The IDF I have in mind is what Smith (2006) has labelled the authorized heritage discourse (AHD), which is the manifestation of a narrative that captures the tangible cultural symbols of the elite social classes, such as 'aesthetically pleasing' sites, monuments and grand buildings. This discourse offers a distinct approach to heritage that has been made to appear factual – thus working to sustain and shape the parameters of social debates regarding heritage issues and the way these represent social relations in both the past and the present. An important product of this relationship is the increasing stability and natural appearance of this current representation of heritage, bringing with it a diminishing of the resources available for resistance (Fairclough 2001a, 33). Paying attention to visual imagery and iconography, and their persuasive repertoire of meaning-making, allows a broadening of the understanding of how things are seen and received.

Authorized Heritage Symbols: Tourism and National Identity

The idea that visual images flow through the cultural sphere, where they are imbued (and themselves imbue) with distinct meanings and values is not new (Jenkins 2003, 307). Of course, visual representations have always been key elements within the subject of tourism and the numerous powerful marketing techniques that knot together this area (Buzinde et al. 2006; Sather-Wagstaff 2008). Likewise, links between tourism and identity have been frequently made in associated literature (see Palmer 1999, 2003, 2005; Henderson 2001; Coleman and Crang 2002). The visuality of tourism, as argued by Zuelow (2005, 189), thus becomes a point of connection at which various ways of seeing – in this case a sense of 'Englishness' – converge in a process of negotiation. For tourism and heritage, this process of negotiation relies upon the communication of identity and belonging, which in many instances has somehow assumed a primordial, or fixed, meaning (Palmer 2003, 428). Institutionalized heritage tourism has thereby become a collection of 'idealized images'; or a mechanism that mediates and constricts the potentiality of the experience (Pritchard and Morgan 2001; Buzinde et al. 2006). It has also become something that extends spatially, culturally and temporally beyond its immediate function.

As Palmer (2003, 428) argues:

> ... 'heritage' is one of the most powerfully imaginative forces due to its association with the notion of historical inheritance. Indeed, the fundamental rationale of heritage tourism is the selection, preservation, and display of nationally significant sites and artefacts designed to promote an idea of nation.

As such, while heritage tourism provides a useful subject area for debating national heritage and identity (Zuelow 2005, 201), it is also a potentially revealing juncture at which to observe the internalization (and operationalization) of a particular IDF. To ignore this interplay and assume that they are simply collections of images used to illustrate the nation's 'sense of itself' is to ignore the discursive factors in operation behind the inclusion/exclusion of images (Zuelow 2005, 189). Take, for example, tourism brochures, which are the standard communicative medium used in this chapter. Brochures produced for the tourism industry effectively package up and symbolize, via a collection of images, a particular understanding of heritage and its own peculiar rendition of 'England's heritage', underpinned as they are by recurrent emphases on aesthetics, monumentality and authenticity. That, it seems, is what 'our heritage' is about.

This distinct sense of heritage, unconsciously shaped by a series of important historical and philosophical contexts, has an undoubted interdiscursive affect on the composition and selection of images for heritage tourism brochures. It is this blurring of boundaries between different social practices in contemporary life that lends itself to the active and continuous articulation of a particular 'way of seeing' (Clarke and Newman 1998, 97; Fairclough 2003, 35). Functioning as part of the apparatus envisioned within Billig's (1995) banal nationalism, heritage tourism brochures present an instance of everyday promotion that holds together a collective national heritage. Moreover, I suggest that this 'way of seeing' has reached discursive closure, so that the authorized heritage discourse, in this instance, is relied upon to provide the images and meanings used to convey what is considered to be 'the nation's communal heritage' (Palmer 1999, 316). The apparent commonsense incorporation of a distinct set of assumptions regarding heritage into touristic brochures is thus revealing of the 'naturalizing tactics' (Thompson 1990, cited in Blommaert 2005, 127, see also Fairclough 1989) and unconscious reproductions of one dominant meaning or vision of heritage. What is at issue, then, is that a limited range of apparently consensual images are pushed forward to define and articulate 'the story of the nation' and its heritage, *at the expense of alternative understandings of heritage*. In the remainder of this chapter, I will explore the relationships I see as existing between the formation and maintenance of the authorized heritage discourse, the visual imagery used in the promotion of heritage, and the circular 'process of closure' (Palmer 2005, 9) this allows in terms of constructions of national identity and 'Englishness' through tourism.

Promoting a People-less Heritage: A Case Study

The responsibility for the promotion of heritage in England falls to the Department for Culture, Media and Sport (DCMS), the governmental department responsible for the tourism industry, English Heritage, the Government's statutory advisor on the historic environment, and Visit Britain, the government sponsored tourism body. Promotional material offered by these organizations form the core of this analysis. Of these, the website Enjoy England,[1] the official site for England's tourism, offers a useful starting point. Within this website, the pages devoted to 'Heritage and Culture' present clear patterns in the types of images chosen to project an image of 'England'. Here, 85 per cent of images (out of a total of 1958) displayed are dominated by abbeys, halls, castles, churches, cathedrals, palaces, ruins, windmills and priories. This cursory exploration allows an immediate glimpse of the persistence of a romanticized, picturesque, ruinous and unchallenging imagery, accompanied by words such as 'historic cities', 'great country houses and gardens', 'Englishness', 'magnificent castles' and 'one of England's finest'. It also provides us with a sense of the limited range of images selected to represent England's heritage, which can be viewed in conjunction with the textual markers used to describe the qualities of that heritage in terms of what 'makes them English'. This presentation of a primarily built environment has an acute affect not only on the idea of England conjured in the tourist imagination, but also strengthens the stronghold of the assumption that England's consensual heritage is hinged upon structurally visible sites, monuments, conservation areas and historic buildings. Importantly, these images present the monological impressions of 'expert' bodies. In so doing, they communicate a certain way of seeing heritage to the wider public, presenting a media-saturated world full of resonant iconic images: devoid of people.

1 Available at: http://www.enjoyengland.com/ (accessed: 26 January 2009).

English Heritage Custodianship Brochures

The images shown in Figure 10.1 represent those commonly found in English Heritage custodianship brochures and wider heritage tourism paraphernalia (see Waterton 2009 for a fuller discussion).

These custodianship brochures are touristic devices designed to advertise English Heritage properties (which total over 400) and can be found at most English Heritage properties and other heritage sites and places, as well as tourist destinations, including bed and breakfasts, tourist information centres, and so forth (English Heritage 2005). As such, access to – and circulation of – the visual images shown within the brochures themselves is very broad, and it is through this ubiquity that the brochures become banal in and of themselves. Each brochure is laid out according to a specific format, with three components arranged to a strict formula: the English Heritage logo adorns the top border, bold and expressive descriptions underlie these, with spectacular pictures commanding the central position. This aesthetically-designed text simultaneously 'represents' and 'advocates' a particular image of heritage, along with the values these images evoke, such as grandeur and monumentality. 'English Heritage', the logo, acts also as a title or label, heading a collective, but generalized appearance of heritage that carries what Bourdieu and Wacquant (2001, 4) term as the 'performative power' to transform into reality that which is being described. 'English Heritage' thus acts to describe and frame England's heritage, becoming a marked and authoritative rhetorical strategy that appeals to the audience to similarly construct and frame heritage in that specific way. In a genre that is explicitly promotional, these three components of logo, description and image establish a particular representation of heritage that is legitimized by the relationship established between them. Together, they are able to communicate a socially encoded message, as with any branding mechanism, which is made more powerful through consensus and repeated formulations of power, fabric and wealth.

The vantage point created by this layout allows the different components to be tied up with each other. The brochure designed for Audley End House and Gardens, *Treasures of Tranquillity*, for example, juxtaposes the image of a 'grand' country house, immaculate and formidable, with ideas of 'treasure' and 'tranquility', thereby expressing characteristics of wealth, worth and composure. This acts as an implicit reference to modality through the evaluative attribution realized through the two nouns used, which comments upon the desirability of the country house. This is, indeed, '[a] national treasure' (English Heritage 2004, 1), a point which is made apparent through the use of extremely transparent markers of evaluation (Fairclough 2003, 173). Different angles are employed by the photographer, affecting the viewers gaze, as are different frame sizes and social distances. For example, Rievaulx Abbey, presented as *Medieval Wealth and Spiritual Vision*, is taken from a height, guiding the observer's eye down towards an aesthetically ordered and carefully managed ruin. It *becomes* a spiritual vision by virtue of this angled vantage point. Likewise, *Power and Valour, Piety and Peace*, a brochure

Figure 10.1 English Heritage Custodianship Brochures (from left to right: Audley End House and Gardens, Treasures and Tranquillity; Explore Historic Northumbria, 16 Great Places to Visit; Historic Yorkshire, Power and Valour, Piety and Peace; Belsay Hall, Castle and Gardens, Indulge in Floral Delights; and Rievaulx Abbey, Medieval Wealth and Spiritual Vision)

Source: With permission of English Heritage

designed to promote Historic Yorkshire, is taken from a vantage point that places Richmond Castle in a position of imposing and considerable power, commanding the surrounding landscape. This angle of imposition is repeated in the brochures for *Explore Historic Northumbria* and *Cumbria and the Lake District*. Another common angle utilized within the promotional material is that of the horizontal angle, which can be used to assess involvement between viewer and object. Kress and van Leeuwen (1999, 390) point to oblique angles as particularly telling of both detachment and involvement, and present an image of something as 'viewed from the sidelines'. In *Treasures and Tranquillity, Medieval Wealth and Spiritual Vision* and *Power and Valour, Piety and Peace*, the images are viewed from an oblique angle, suggesting that the viewer is not involved with what they are viewing. Rather, the choice of this angling signals a disjuncture between this world and our world.

Likewise, the social distance set up between the objects portrayed within the images and the viewer communicate the boundaries assumed to exist between them, and the social distance drawn upon to configure those boundaries. Hall (1964, cited in Kress and van Leeuwen 1999, 387) refers to these fields of vision as the 'proxemics' of everyday interaction. The images used both on the front covers of each promotional brochure, and within the brochure itself, occupy positions of 'public distance' or 'far social distance', which indicate relationships which at best are formal and impersonal, and at worst, as existing between '... people who are and are to remain strangers' (Kress and van Leeuwen 1999, 387). In conjunction with this social distance and angling of involvement, an image of heritage is utilized that carries nothing of the banality of the everyday or 'he familiar' quite the contrary, these images radiate with 'exceptional', 'powerful', 'excessive' and 'luxurious' ideas of significance. The message: one does not 'indulge' in the ordinary. Instead, heritage becomes something to bask in, a detour from the familiar, in a discursive construction explicitly created by the brochures. This message, however, is also one of contradiction, in which the 'exceptional' and 'extraordinary' are marketed and promoted with a reliance on ubiquity and its attendant banality. This is witnessed by the vast array of trivial signs and cultural forms associated with national heritage tourist sites.

While there are a number of ways by which to approach heritage, these directions are not accommodated in the focus that dominates the images, which clearly identifies the fabric of heritage as paramount, defined along lines of monumentality, aesthetics and the grand. This emphasis is established visually by the positioning of the images and the objects within them – the eye is drawn to the structures, which are centrally placed so as to dominate. The soft tints and mix of colours also capture a sense of mystery, symmetry and promise, all of which add to the beautification of the past. Importantly, the images also communicate distance and age, generating a past that is external and does not demand that the audience enter into a subjective relationship with it (Kress and van Leeuwen 1990, 28). Indeed, the relationship between the heritage site or place and the tourist, user or audience becomes not so much about connecting 'with place', but with making an idea of something – in this case heritage – real, substantial or complete

(MacCannell 1999, 112). The weight and 'worth' expressed by the images and accompanying inscriptions therefore engage in a process of signalling themselves as markers of meaningful experience capable of making this reality (MacCannell 1999, 112). Not only, then, do the brochures rest on images of consensus; they also draw heavily on ideas of disconnection, which is profoundly at odds with ideas of belonging and identity.

It is here that a discourse dealing explicitly with an innate and immutable sense of value can be discovered, which commands attention be placed on the fabric of heritage itself, as opposed to any associations and constructions of identity that may be ascribed to that fabric. In this sense, the visual imagery used by the promotional brochures slices off the deeper understandings of heritage in favour of an assumed universal significance that is seen to exist within that place or aspect of material culture. From this we see a generalized and singular heritage that works to suppress any aspirations of local and regional identity. What becomes apparent is an understanding of heritage and identity that is firmly drawn along the lines of similarity, rather than difference, allowing the unity of people considered in the management process to be defined specifically in terms of good, educational and conflict-free. In short, this is reflected in the extraordinarily high commitment to one image of heritage, which can arguably act as a marker of a categorical, non-modalised assertion: heritage is fabric, monumental and grand. This brings the argument back to earlier remarks regarding commitments to future generations. Essentially, utilizing a particular representation of heritage leaves little room for other voices to be heard, and the presence of present generations hardly felt. In these representations, the role of people in the management process has been re-contextualized and layered with indifference, such that the only element that is consistently prominent is the materiality itself. This dialogically closed assumption sits far more comfortably with a favouring of the rights of future generations or the unborn (who, essentially, will never come).

The 2004 Member's and Visitor's Handbook

The idea that heritage belongs to English Heritage ('our' sites and monuments) – and is somehow simultaneously people-less and an exclusive family space – is perhaps most explicitly expressed in the English Heritage Member's and Visitor's Handbooks (editions 1999/2000, 2003/2004 and 2007/2008) (see Figure 10.2), which outline what English Heritage (2007, 3) has '... to offer you and your family'.

Drawing on a perspective that is reminiscent of the great and the masterful, the images included in each promotional handbook are managed around a perception of heritage that is reflective of the traditional narratives of history and the past. The handbooks are generally a 250–300 page collection of English Heritage properties that are clearly promotional in genre. Produced by English Heritage under the direction of both an Art Director and Picture Director (English Heritage 2004, 5),

Figure 10.2 English Heritage 2004 Member's and Visitor's Handbook (top left: Old Wardour Castle; top right: Royal Garrison Church; and centre: Pendennis Castle)

Source: With permission of English Heritage

they are designed to endorse and communicate the appeal of England's historic environment both internationally and domestically. The handbook comprises part of the English Heritage membership package, and is thus freely accessible to all members, of which there are around 550,000 (English Heritage Customer

Services, pers. comm., June 2005). It is also available for purchase by non-members. While the Handbook is divided into regions, each summarized in terms of its historic country houses, gardens, parks, statues, ecclesiastical buildings, sites, windmills, castles and monuments, the similarity of both format and images across the spread of handbooks is striking. Each property listing includes a brief description, details of opening times, their family orientation and information regarding access, transportation and entry cost. Many, but not all, of the properties are also accompanied by a visual image.

Across this visual imagery, monumentality is central. Images depict a glamorous, stunning, mysterious, masterful, dramatic, beautiful and historic national and regional heritage. The use of colour, lighting and other enhancing techniques is impressive, resulting in a glossy handbook that commands admiration and depicts a national history full of wealth, extravagance and architectural merit. Indeed, these are precisely the type of characteristics one would expect a promotional genre to draw upon. The images included are arranged with artistic flair, and together make an advertising package evocatively displaying aerial photographs, snap-shots of forts, castles and other imposing buildings nestled amongst picturesque landscapes and close-up images of particularly impressive features.

A defining feature of each handbook is the staged appearance of the visual images. The visual design is invariably composed around a number of interesting features that create axes of sight. For example, the cannon in the foreground of the left picture in Figure 10.2, Pendennis Castle, acts as a frame for the castle, both in terms of a historical 'prop' and a horizontal axis that directs the viewer's eye up towards it. Utilizing Kress, Leite-Garcia and van Leeuwen's (1997) division of images between bottom and top, real and ideal allows this image to revolve very closely around the horizontal axis, moving the image of the 'ideal' castle into the centre of attention, prominent for all points of view. Yet at the same time, the discursive ideal of the castle is distant, and somewhat marginal and mystified. This coded distance positions the reader of the Handbook as a viewer, and therefore unengaged with the image. One Gun Battery takes up the position of 'real' in the immediate foreground, providing an element of factualness against which to compare the presence of an exaggerated 'ideal' offered by the castle.

The remaining images in Figure 10.2, Old Wardour Castle and Royal Garrison Church, are illustrative of another important feature of England's heritage in terms of both identity and tourism. Old Wardour Castle is taken with a skyscan balloon, and presents an image of a romantic castle ruin set amongst finely mown lawns and designed landscapes. The Royal Garrison Church image offers a startling contrast between blue skies and the fabric of the building, with sharp angles, clean shapes and an overall sense of precision dominating. In both images, what radiates is a sense of the 'cared for', immaculate and neat management of heritage, even when in ruins, and it is this that becomes that important part of demonstrating what 'managing the past' is all about. Caught up in their own chain of events, the images are articulating what function preserving the past performs for developing a sense of Englishness and national identity. It is in the 'cared for' images that heritage is actively doing something: it is creating a sense not only of what heritage is, but what it means to manage that heritage (Emerick 2003, 113). This is reminiscent of an argument developed by Schwyzer (1999, 58), who suggests, with reference to the White Horse of Uffington, that it is '... upkeep itself, not what is being kept up, that expresses the spirit of the nation.' With reference to the Elgin Marbles, he goes on to argue that:

> The Elgin Marbles may be foreign in origin, but only England can be trusted with their preservation – an activity so deeply English that it transforms these ancient Athenian artworks into a feature of the national heritage (Schwyzer 1999, 58).

Following from Schwyzer, heritage can thus be seen to be attached to a distinct sense of national identity, which is reinforced and re-materialized through tourism and marketing strategies. This identity takes precedence over a number of other possibilities, directing a conversation that is styled around communicative strategies between those managing heritage and its customers or consumers, rather than stakeholders. The authorized heritage discourse as an IDF is visible in this instance through its appeals to preservation, conservation and presentation of an aesthetically pleasing site, building or monument. Idealized views of heritage have thus been captured in the glossy photos, again with little or no interaction demonstrated between people and place.

Discussion

The social world of heritage is thus depicted in terms of abstraction, generalization and preservation, allowing the complex issues that surround the heritage management process, including the various stakeholders and interest groups, to be removed. Fairclough (2003, 138) argues that when things are being generalized it is necessary to question what classification schemes constitute this particular vision. In terms of the case studies provided in this chapter, we can see that visions of dominance, grandeur and a certain sense of aesthetics take precedence, alongside the stunning imagery of haunting ruins, powerful buildings and indulgent wealth. With this, the authorized heritage discourse is able to recontextualise what heritage is along the lines of the consistently present built fabric. The various images gathered for this analysis thus become part of a series, moving together from the same vantage point towards the same end message. Moreover, the authors of the touristic brochures, by presenting heritage in this way, assert an image of ultimate control over heritage and its' management, and by implication, peoples ideas of what constitutes heritage and indeed the past. It is thus an image of alienation. By textually applying an emphasis on the pronoun 'our', and then perpetuating a divide through the static illustrations of England's monumental heritage, the leaflets, handbook and websites rebound with questions of the noticeable distinction implored between 'the heritage experts' and 'the wider public'. Clearly, possession of heritage is placed outside of the wider public, and likewise, decisions concerning its management and significance.

Produced between 2000 and 2007, these representations should actively illustrate inconsistency and conflict as various heritage institutions attempt to come to terms with the social inclusion agendas and policies of the Labour government. Yet, they persistently work to support a discourse that emerged in a timeframe dominated by Romanticism and enlightenment sentiments, thereby signalling the extent to which the effects of 'truth' have naturalized a very particular 'way of seeing'. The absence of contradiction, complexity or even variety within the images signifies the depths of stability held by this discourse, and the diminished resources available for alternative approaches. Not only does this stability and consensus hold across ideas of heritage, it holds, too, across tourism literature, signifying and sustaining a very particular idea of England's national identity. By virtue of this coupling of tourism and heritage, a single way of seeing has found domestic, national and international acceptance and resonance. What is particularly important about this relationship is the discursive circularity it incites. Here, as McLean and Cooke (2003, 155) point out, many tourists begin to see an imagined English identity in the distinctive markers utilized, while simultaneously, it must be remembered that '... the way in which we see ourselves is substantially determined by the way in which we are seen by others' (O'Connor 1993, 68). As such, this circularity, itself, is working to naturalize and sustain the AHD. In many ways, this circularity is then aided and sustained by wider marketing imperatives, which latch onto – and promote – particular aspects of heritage and their narratives. Tourism, through avenues of promotion and marketing, thus becomes, as Dicks (2000, 174) points out, a potent force to be reckoned with, or a 'technology of power' capable of defining and selecting heritage for a range of user groups.

In terms of the publications, the series offered here forms part of what can be labelled as a generic chain (Fairclough 2001b, 255), in which different elements of the chain can be assumed to frame the others. Indeed, all are part of a larger generic chain that includes other promotional documentation, webpages, policy documents, press reports and debates. Communication is thus characterized here in the responses visible in and between the links in this generic chain. While the persuasive devices used to promote a particular idea of heritage in all of the examples are not limited to the same mix of genres, the imagery in all three can be said to be performing in a 'reader directed' capacity. The images play a promotional role. They illustrate the grandeur, power and aesthetically pleasing characteristics of England's heritage – and these are characteristics that are presented as unquestioned and stable, with any alternative voices concerning heritage are absented. Significantly, any arguments developing between old and new ways of thinking about heritage and its management in wider political and theoretical spheres are not reflected here. They are not asking 'what' and they are not asking 'why'. That argument is apparently over.

What these images ignore, then, is the large number of representations jostling for position within the heritage management process: they ignore the interplay of people with heritage, and the resultant conflict over meaning. Therefore, the representations – and both producers and audiences – reflect an absence of what discourse-analytical theorists have labelled the 'argumentative texture' (see for example the edited collections of Fischer and Forester 1996; Hajer and Wagenaar 2003; Wetherell et al. 2001). An important aspect of this texture is the interaction between competing, and often contradictory, viewpoints, which may lead to both contestation and shifts in the relations of power surrounding an order of discourse (Wetherell 2001, 25). This is of particular interest when one particular representation emerges out of the argumentative texture and begins to mobilize meaning (Wetherell 2001, 25). It is this point that is useful for reflection when considering the use of imagery by heritage organizations, particularly the aim of the leaflets and brochures, to reach a wide number of publics within their audience (van Dijk 2001, 309). The emergence of a singular approach to what constitutes heritage signals the materialization and maintenance of a dominant discourse, in this case, one bound up with a static, built environment (see Watson this volume).

Quite naturally, iconography that leaves out the interaction of people and heritage, thereby downplaying the intersection of identity, affects the authority of these brochures. A large number of people are exposed to this medium of communication, and the assumptions they contain. Further, as these are contained within an authoritative act, their effectiveness is maximized (van Dijk 2001, 309), and carry the potential to become sites of alienation or exclusion (Torfing 1999, 211). Few who see these images could begin to draw in an understanding of the meanings these may carry in reality for local communities. Indeed, the question may be raised as to how English heritage/tourism organizations may be expected to promote anything beyond a national heritage, but surely the question of how the meanings connected to heritage can be sustained is of central importance to those management bodies. Essentially, what is triggered by recent introductions of social inclusion objectives and community foci is a significant destabilization of the 'national' focus, and this is particularly relevant to English heritage. Sharon Sullivan, former director of the Australian Heritage Commission (AHC), in different circumstances, made the statement that 'management is only effective if it is rooted in the values of the culture whose heritage is being managed' (cited in McBryde 1995, 8). It therefore stands as a striking paradox that heritage is illustrated as people-less, and is authoritatively demonstrated to 'the public' in a dialogically limited expression.

Conclusion

This chapter focuses on revealing the discursively constructed terrain of meaning that underpins the seemingly unproblematic institutional signification of

English Heritage, DCMS and other bodies, and situates heritage tourism within a management process replete with questions, conflicts and debate – indeed, it carries a noticeable argumentative texture. Despite this, a particular image or vision of heritage is consistently put forward, which is suggestive of an underlying assumption that sees heritage as structurally visible sites, monuments and buildings that are both monumental and imbued with extraordinary significance. Moreover, this image of heritage ignores the vital dimension of people themselves. But what does this information actually do for us? What possibilities lie in the conclusions of critical discourse-analytical research? Principally, the emerging possibilities draw from the understanding that to comprehend the social life of heritage, one must ultimately deal with language, albeit in this instance visual language. As argued by Fairclough (2003, 203), this is because language has become more salient: 'more important than it used to be, and in fact a crucial aspect of the social transformations which are going on – one cannot make sense of them without thinking about language'. Indeed, this argument must incorporate the potency of the visual, which does not act merely to describe or depict, but is used to accomplish a range of intended, but also in many cases unintended or unrecognized, objectives. The most striking point to emerge here, then, is the apparent lack of a need to deconstruct visual imagery dealing with heritage, despite the pervasive currency it carries. Instead, visual imagery becomes uncomfortably under-scrutinized, and holds questionable modality: through this lack of clarity, the images emerge as rhetorical tools that are able to manage and define clear cut distinctions between what heritage is and what heritage is not. The calculated selection of particular images describes, invokes and therefore advocates a particular idea of heritage that carries a performative power (Bourdieu and Wacquant 2001, 4) capable of simultaneously creating, in a sense, that to which it is referring.

Acknowledgements

The research providing the basis for this contribution was generously funded by the Arts and Humanities Research Council (AHRC) and the Research Councils United Kingdom (RCUK). I am also grateful to Dr Keith Emerick and Dr Robin Wooffitt for their helpful comments on an earlier version of this chapter. Finally, thanks are due to Dr Laurajane Smith and Gary Campbell for reading and commenting upon all drafts of this contribution.

References

Benwell, B. and Stokoe, E. (2006), *Discourse and Identity.* (Edinburgh: Edinburgh University Press).

Billig, M. (1995), *Banal Nationalism.* (London: Sage Publications).

Blommaert, J. (2005), *Discourse: A Critical Introduction.* (Cambridge: Cambridge University Press).

Bourdieu, P. and Wacquant, L. (2001), 'NewLiberalSpeak: Notes on the New Planetary Vulgate', [Online]. Available at: http://www.radicalphilosophy.com/default.asp?channel_id=2187&editorial_id=9956 [accessed: 15 June 2005].

Buzinde, C.N., Santos, C.A. and Smith, S.L.J. (2006), 'Ethnic Representations: Destination Imagery', *Annals of Tourism Research* 33:3, 707–28.

Clarke, J. and Newman, J. (1998), *A Modern British People? New Labour and the Reconstruction of Social Welfare.* Paper to the Discourse Analysis and Social Research Conference, Ringsted, Denmark, 24–26 September 1998.

Coleman, S. and Crang, M. (2002), 'Grounded Tourists, Travelling Theory', in S. Coleman and M. Crang (eds), *Tourism: Between Place and Performance*, 1–17. (Oxford: Berghahn Books).

Crouch, D. and Lübbren, N. (2003), *Visual Culture and Tourism.* (Oxford: Berg).

Dicks, B. (2000), *Heritage, Place and Community.* (Cardiff: University of Wales Press).

Durrheim, K. and Dixon, J. (2001), 'The Role of Place and Metaphor in Racial Exclusion: South Africa's Beaches as Sites of Shifting Racialization', *Ethnic and Racial Studies* 24:3, 433–50.

Emerick, K. (2003), 'From Frozen Monuments to Fluid Landscapes: The Conservation and Preservation of Ancient Monuments from 1882 to the Present', unpublished PhD thesis, University of York.

Emmison, M. and Smith, P. (2000), *Researching the Visual: Images, Objects, Contexts and Interactions in Social and Cultural Inquiry.* (London: Sage Publications).

English Heritage (2004), *2004 Members' and Visitors' Handbook.* (London: John Brown Curtis Publishing).

English Heritage (2005), Homepage. [Online]. Available at: http://www.english-heritage.org.uk [accessed: 19 June 2005].

English Heritage (2007), *2007/8 Members' and Visitors' Handbook.* (London: John Brown Curtis Publishing).

Fairclough, N. (1989), *Language and Power.* (London: Longman).

Fairclough, N. (1995), *Critical Discourse Analysis: The Critical Study of Language.* (Essex: Pearson Education Limited).

Fairclough, N. (2001a), 'Critical Discourse Analysis', in A. McHoul and M. Rapley (eds), *How to Analyse Text in Institutional Settings: A Casebook of Methods*, 25–38. (London: Continuum).

Fairclough, N. (2001b), 'The Discourse of New Labour: Critical Discourse Analysis', in M. Wetherell, S. Taylor and S.J. Yates (eds), *Discourse as Data: A Guide for Analysis*, 229–66. (London: Sage Publications).

Fairclough, N. (2003), *Analysing Discourse: Textual Analysis for Social Research.* (London: Routledge).

Fischer, F. and Forester, J. (eds) (1996), *The Argumentative Turn in Policy Analysis and Planning.* (Durham: Duke University Press).

Hajer, M. and Wagenaar, H. (eds) (2003), *Deliberative Policy Analysis: Understanding Governance in the Network Society.* (Cambridge: Cambridge University Press).

Hall, S. (2005), 'Whose Heritage? Un-settling "the Heritage", Re-imagining the Post-nation', in J. Littler and R. Naidoo (eds), *The Politics of Heritage: The Legacies of 'Race'*, 23–35. (London: Routledge).

Henderson, J. (2001), 'Heritage, Identity and Tourism in Hong Kong', *International Journal of Heritage Studies* 7:3, 219–35.

Jaworski, A. and Coupland, N. (1999), 'Introduction: Perspectives on Discourse Analysis', in A. Jaworski and N. Coupland (eds), *The Discourse Reader*, 1–44. (London: Routledge).

Jenkins, O.H. (2003), 'Photography and Travel Brochures: The Circle of Representation', *Tourism Geographies* 5:3, 305–28.

Kress, G. and van Leeuwen, T. (1990), *Reading Images.* (Victoria: Deakin University Press).

Kress, G. and van Leeuwen, T. (1999), 'Representation and Interaction: Designing the Position of the Viewer', in A. Jaworski and N. Coupland (eds), *The Discourse Reader*, 377–404. (London: Routledge).

Kress, G. and van Leeuwen, T. (2006), *Reading Images: The Grammar of Visual Design.* (London: Routledge).

Kress, G., Leite-Garcia, R. and van Leeuwen, T. (1997), 'Discourse Semiotics', in T.A. van Dijk (ed), *Discourse as Structure and Process*, 257–91. (London: Sage Publications).

MacCannell, D. (1976[1999]), *The Tourist: A New Theory of the Leisure Class.* (Berkley: University of California Press).

McBryde, I. (1995), 'Dream the Impossible Dream? Shared Heritage, Shared Values, or Shared Understandings of Disparate Values?' *Historic Environment* 11:2/3, 8–13.

McLean, F. and Cooke, S. (2003), 'Constructing the Identity of a Nation: The Tourist Gaze at the Museum of Scotland', *Tourism, Culture and Communication* 4, 153–62.

Morgan, N. and Pritchard, A. (1998), *Tourism Promotion and Power: Creating Images, Creating Identities.* (Chichester: Wiley).

O'Connor, B. (1993), 'Myths and Mirrors: Tourist Images and National Identity', in B. O'Connor and M. Cronin (eds), *Tourism in Ireland: A Critical Analysis*, 68–85. (Cork: Cork University Press).

Palmer, C. (1999), 'Tourism and Symbols of Identity', *Tourism Management* 20, 313–21.

Palmer, C. (2003), 'Touring Churchill's England: Rituals of Kinship and Belonging', *Annals of Tourism Research* 30:2, 426–45.

Palmer, C. (2005), 'An Ethnography of Englishness: Experiencing Identity Through Tourism', *Annals of Tourism Research* 32:1, 7–27.

Pritchard, A. and Morgan, N.J. (2001), 'Culture, Identity and Tourism Representation: Marketing Cymru or Wales?' *Tourism Management* 22, 167–79.

Rose, G. (2001), *Visual Methodologies: An Introduction to the Interpretation of Visual Materials.* (London: Sage Publications).
Sather-Wagstaff, J. (2008), 'Picturing Experience: A Tourist-centered Perspective on Commemorative Historical Sites', *Tourist Studies* 8:1, 77–103.
Schwyzer, P. (1999), 'The Scouring of the White Horse: Archaeology, Identity, and "Heritage"', *Representations* 65, 42–62.
Smith, L. (2006), *Uses of Heritage.* (London: Routledge).
Torfing, J. (1999), *New Theories of Discourse: Laclau, Mouffe and Žiže.* (Blackwell Publishers: Oxford).
Urry, J. (1994), 'Europe, Tourism and the Nation State', in P.C. Cooper and A. Lockwood (eds), *Progress in Tourism: Recreation and Hospitality Management Research*, 89–98. (Chichester: Wiley).
van Dijk, T. (2001), 'Principles of Critical Discourse Analysis', in M. Wetherell, S. Taylor and S.J. Yates (eds), *Discourse Theory and Practice: A Reader*, 300–17. (London: Sage Publications).
VisitBritain (2004), *Heritage Britain.* (London: VisitBritain).
Waterton, E. (2009), 'Sights of Sites: Picturing Heritage, Power and Exclusion', *Journal of Heritage Tourism* 4:1, 37–56.
Wetherell, M. (2001), 'Themes in Discourse Research: The Case of Diana', in M. Wetherell, S. Taylor and S.J. Yates (eds), *Discourse Theory and Practice: A Reader*, 14–28. (London: Sage Publications).
Wetherell, M., Taylor, S. and Yates, S.J. (eds) (2001), *Discourse Theory and Practice: A Reader.* (London: Sage Publications).
Zuelow, E.G.E. (2005), 'The Tourism Nexus: National Identity and the Meanings of Tourism Since the Irish Civil War' in M. McCarthy (ed), *Ireland's Heritages: Critical Perspectives on Memory and Identity*, 189–204. (Aldershot: Ashgate Publishing).

Chapter 11
Time Machines and Space Craft: Navigating the Spaces of Heritage Tourism Performance

Tom Mordue

The last three decades have produced significant changes in what Harvey (1989a) calls 'the urban process'. Many Western cities have been restructured as places of consumption as well as places to live and work in which cultural and heritage assets are show-cased and turned into commercial products that are integrated into the fabric of central public spaces. Urban centres are also subjected to branding strategies aimed at attracting tourists while promising that a particular town or city will provide a uniquely exquisite location in which to do business and a 'quality of life' that meets service class workers 'lifestyle' aspirations. Initiatives such as these are seen in the urban studies literature (Meethan 1997; Atkinson 2003; Bayliss 2004) as particularly interesting features of many post-industrial cities' economic development aspirations, where, because of 'the collapse of the industrial base of their cities and the rise of the service sector, city decision-makers [have] prioritized economic development and turned to the arts and culture as one area with considerable potential' (Bayliss 2004, 818). Not only post-industrial cities are affected; many 'heritage cities' (Urry 1995) are subjected to the same commercial forces and are turned into themed 'time machines' that offer heritage experiences for consumers by urban managers keen to attract a diversity of investment capital while fending off increased competition from a plethora of cultural spaces opening up to the 'tourist gaze' (Urry 2002).

These seemingly universal trends are not just a matter of globalization visiting remorseless homogenization and commodification upon varied people, places, cultures and heritage, but are outcomes of complex negotiations between the global and the local that involve 'institutional elements as much as the micropolitics of daily experiences' (Meethan 1997, 334; after Healey et al. 1995). The main filter of mediation happens at the level of urban governance because this is the formal decision-making switch-point between the local and the global – though the consequences of decisions made here will be felt most tangibly at street level. It is, therefore, important to understand how this mediation process affects the spaces of everyday life that are subject to 'urban renaissance' (Bianchini 1990) strategies and spectacularizations as heritage tourist attractions. This chapter broaches this issue in an urban tourism context by coupling an analysis of how urban governance has changed in recent years with a critical assessment of how the current mixing of commerce and culture makes new demands on the nature, use and meaning of urban public space for 'everyday' people and practices. Visually, public spaces are being re-aestheticised to fit the 'gaze' of tourists who are increasingly mobile and are thus able to compare the attributes of one location with many others. Similarly, the everyday environs where local people once lived their local lives are recast and pitched into global tourist space, making local people who enter these spaces objects of the gaze and cast members in global heritage tourism performance.

By drawing upon a range of research undertaken in York city centre in North East England, these critical insights are grounded to provide an illustrative example of how city centre space is crafted, visually constructed, contested and reproduced by processes of heritage tourism development and urban renaissance, and by 'the public' who live and consume it through street-level interactions.

Performing, Regulating and Globalizing Tourist Space

Tourist sites are especially rich performative arenas where the organization and interpretation of space provides a framework within which locals and visitors ascribe and contest meaning (Chaney 2002). As Chaney points out, locals and visitors do not constitute homogenous categories; the meaning of place for them will be dispersed, exist on a number of dimensions and will change according to differing expectations. Furthermore, Griffiths (1993) shows that cultural consumption in city centres is a local reflection of wider socioeconomic power, which can create a 'dual city' where the middle-classes are elevated into the socio-spatial core, while lower social classes are exiled to the socio-spatial periphery because they have neither the cultural nor economic capital to be significant consumers of culture. Thus urban heritage space is not only crafted, themed, and timed by heritage tourism brokers and producers, but contested by the various constituent groups who perform it on a daily as well as strategic basis.

The use of performance as a metaphor for tourist practice has become a critical focus of attention in the tourism literature in recent years (see for example Edensor 2000, 2001; Chaney 2002; Coleman and Crang 2002; Bagnall 2003; Smith 2006). Yet considering how local people encode and enact performances that compare and compete with those of tourists who occupy the same space is a relatively new vein of analysis (Mordue 2005). Although these are embodied performances that engage all the senses as people move through space, they are also discursive (cf. Tulloch 1999). This is because through narrative people author what they and others do to create and contest places as 'performative events' (Coleman and Crang 2002). In urban tourism settings, performative events tend to take place in what Edensor (2000) calls 'enclavic tourist space', spaces that are carefully crafted and manicured for cultural/heritage consumption. Like Griffiths (1993), Edensor also alludes to how social exclusion is commonplace in tourism enclaves because classificatory struggles produce discourses that rise to prominence to author what is appropriate activity within their boundaries, within which '"[u]ndesirable elements" and social practices … are likely to be deterred' (2000, 328). In urban heritage space it is thus common that 'aesthetic appreciation of particular landscape styles and patterns of consumption can be subtly defined and redefined to exclude others' (Duncan and Duncan 1997, 170; see also Duncan 1999). Likewise, the official designation of landmark historic districts are 'often influenced by which social groups who will consume [them] … moreover, a landmark designation creates both monopoly rents and a monopoly of consumer rights' (Zukin 1990, 42).

The resulting visual effect or 'spatial narrative' is constructed around 'a symbolic quest for authenticity, validation and monumentality, as well as a myth that a historically preserved enclave – and others like it – represent the real, historical city' (Zukin 1990, 42). Furthermore, historic enclaves confer a sense of place, and therefore a sense of authenticity, to the things being sold within them (Halewood and Hannam 2001). They are subject to what I am referring to as 'space craft', where their landscapes are cleaned up, visually enhanced and extensively signed, let out at high rents to businesses and other tenants. As a consequence, property values rise and these landscapes are generally turned into consumption spaces that take full advantage of the cultural and economic synergies between heritage, tourism, shopping and gentrification. Through this, the cultural capital of place is wedded to the habitus of new middle-class groups (Zukin 1990) who are particularly prominent in the creation and consumption of culture and heritage (see also Tunbridge and Ashworth 2000).

While these issues may be played out locally, they indicate the global nature of tourism development and consumption. Moreover, in Western heritage tourism there is a recursive tendency to follow a formulaic norm that reproduces a visual aesthetic of 'pastness' in urban space (Corner and Harvey 1991). Thus, a significant amount of space craft, or spatial design and management, is used to entice consumption. In business parlance, then, cities become 'products' that need to perform in the global market, consisting of 'resources' by which attractions considered part of the authenticity and uniqueness of place are often mixed with other more 'exotic' attractions that have blanket appeal to cosmopolitan consumers. The result is a de-differentiation of the local and global in which local particularity becomes disembedded, or, as Robins (1991, 31) puts it, 'torn out of time and place to be repackaged for the world bazaar'. Thus, when 'global tourists' seek place distinction, they are complicit in a paradoxical effacement of local difference by the tourism industry, because in the competition for tourists places simultaneously offer uniqueness and global standards of service delivery. This tourism 'glocalization' both ensures the 'quality' of the 'product' in universal terms and compels the staging of place particularities and local cultures as attractions.

Urban Renaissance and the New Urban Governance

Political economy theorists have situated such developments in urban centres within the shifts of Fordism to post-Fordism (Harvey 1989a,b; Lipietz 1993). A main tenet of this thesis is that increased competition between places for footloose global capital has, since the early 1970s, brought more entrepreneurial forms of urban governance to virtually all Western cities. US cities have been trend-setters in this, which Molotch (1976) described as 'growth machines' that would form 'growth coalitions' between private and public agencies to attract investment and consumption capital into tired downtown spaces. Typically, synergies between culture, heritage, retailing, tourism, property development and restoration would be sought to gain competitive advantage, although beneficiaries were ostensibly the cities' powerful elites rather than their communities at large (Molocth 1976; Logan and Molotch 1987; Madrigal 1995). Pittsburgh and Baltimore are often cited examples of how declining industrial economies were subject to what became known in the US as 'boosterism' based on place promotion and spectacular urban regeneration (see Harvey 1989a,b; Griffiths 1993).

This consumerist/promotional model of urban renaissance also became dominant, though by no means universal, in Europe (Griffiths 1993). Because many European cities have long histories that are visibly expressed in their buildings and landscapes, heritage has become central in many of their development trajectories. As such, these cities have become themed 'time machines' that facilitate boosterism and growth – where culture, commerce and urban regeneration are conflated – to produce the urban growth machines' that Molotch (1976) alerted us to more than 30 years ago. In 1980s Britain, Glasgow was the first notable adoptee of this type of urban regeneration, although many other post-industrial British cities pursued 'boosterist' like strategies at this time (Bianchini 1991). The momentum gathered pace in Britain during the 1990s, notably in cities controlled by previously resistant Labour-led councils (Hughes 1999). Since then 'new urban governance' has emerged, which is about:

> ... Working across boundaries within the public sector or between the public sector and the private sector or voluntary sectors. It focuses attention on a set of actors that are drawn from but also beyond the formal institutions of government. A key concern is processes of networking and partnership (Stoker 2000, 3).

Integral to this are discourses and practices of 'new public management', advocating management strategies adopted by the private sector in order to create 'modernized', efficient and effective public institutions (Astleiner and Hamedinger 2003). Astleneir and Hamedinger (2003, 54), however, like Hughes (1999), warn that this movement sees people 'as consumers rather than citizens', and question whether it represents 'a de facto downgrading of the role of the public'. Thus, new urban governance does not only represent structural shifts in the way cities are managed in the name of a previously 'failed' public but is also about new ways of addressing, thinking, knowing, interpreting and disseminating what 'the public' is, what is in its interest and how it should be 'served'. What this might mean for the status of urban public space is a salient issue, especially given that what constitutes boundaries between the state, the market and civil society are being challenged.

Discussing the growing significance of tourism to urban development and regeneration, Shaw and Williams (2004, 206) note 'that national and local governments have increasingly sought to build [public/private] partnerships for local development that are focused on, or incorporate, tourism'. They point out that this 'new entrepreneurialism' brings benefits regarding increasing the capacity to tackle complex development problems because traditional government boundaries, interests and response capabilities are not sufficiently attuned to, and do not easily map onto, the multi-complexity of underlying social processes. However, they echo other authors' concerns by arguing that these types of initiative result in a troublesome 'blurring of the boundaries between the public and the private' (Shaw and Williams 2004, 206). The problematic here is that rather than transcending inequalities in society, tourism partnerships could reinforce them by representing the interests of the most powerful partners more effectively. The payback for private interests, then, would be greater access to street-level interactions and the legitimization of their own priorities as being synonymous with 'the public good' or the greater interests of the city in which the partnership is located. From such a perspective, partnerships are not radically new in that they have greater potentialities to deliver more effectively democratic urban development, but are effective in their ability to reach into everyday urban life while maintaining the power of strategic focus.

Socio-spatial Governance, Tourism, Leisure and Democracy

This panoptic and invasive capacity of partnerships invites some discussion of what Foucault called 'governmentality', which is about how government power is increased through the creation of specific types of institution, knowledge and experts that fashion a particular subjectivity in individual citizens. Urban partnerships, then, can be thought of as instruments of a disciplinary gaze trained on normalizing people's everyday interrelations. Moreover, through routine and habitual quiescence, urban citizens internalize this gaze to become self-managed subjects of the state (Foucault 1991, 1997; Bridge and Watson 2002). The corporeality of people's spatial interactions is what makes such surveillance so effective because the threat of being seen is ever present. Thus, power is embedded in institutions and, with panopticism, is mobilized through socio-spatial practices. This, however, is not a one way street. Individuals offer multiple points of resistance and micro-level strategies to combat such omnipotence, because the mind is less available to normalization than the body, and, anyway, the capillaries of the social are many and complex (Thrift 2004).

Urry adapted a Foucauldian perspective in his analysis of how 'the tourist gaze' is socially produced, organized and systematized through the actions of 'many professional experts who help construct and develop our gaze as tourists' (2002, 1). Assessing how this manifests spatially, Edensor (2000) describes how local authorities, in conjunction with powerful commercial interests, develop 'tourist enclaves' that are subjected to high degrees of spatial regulation. So often found in urban settings, tourist enclaves are purified spaces that are visually, aesthetically and socially set apart from everyday life, where '"undesirable elements" and social practices are likely to be deterred' (Edensor 2000, 328). Given this, and given that a kind of specialized visuality based on street entertainments, tourist shopping, hotels, restaurants, various cultural attractions and tourist signage focus the urban tourist gaze in particular ways, spontaneous social contact is likely to be minimal in the enclave. How inclusively 'public' these tourist enclaves are is thus an important political issue, although it is too simplistic to assume that once tourism comes to town public space is given over to a wholly oppositional 'tourist space' (MacCannell 1976).

In practice, the public space/tourist space binary is refuted by the many public officials and private investors who voice the familiar clarion call that urban tourism development benefits the local economy and physical environment while providing greater leisure and 'lifestyle' opportunities for the local populace as a whole (Hall 1994; Madrigal 1995; Hannigan 1998; Urry 2002). Although a contentious political message, this does signal how issues surrounding local leisure and global tourism are increasingly intertwined. Moreover, because '[t]ourism is a leisure activity which presupposes its opposite, namely regulated and organized work' (Urry 2002, 2), tourist enclaves are constructed as performatively distant and visually distinct from the everyday spaces of workaday life, whether they are physically proximate or afar in the home environments of tourists. In this way, the enclave is discrete yet de-differentiated and 'glocalized', and can be variously sold to high value tourists, locals, new investors and incoming key workers as a cultural attraction, a leisure facility, a lifestyle attribute, a good place to do business, and a public good.

Degen (2003, 879) argues that such spatial orchestration means that ethical and social issues inevitably take a back seat to aesthetic goals in the contemporary city, where:

> ... [n]ew public spaces are emerging that are fostering new forms of public life, qualitatively different from modernity. Exclusion or inclusion in these spaces is fostered through the sensuous regimes in the place, and the imposition and control of new practices are often disguised as leisure or culture.

Despite the façade of being enhanced public goods, entertainment, a distinct visuality and spectacular consumption are subsuming sociality and democratic citizenship in these places. Furthermore, by altering the existential and sensed experience of place through aesthetic management, certain groups, notably lower class groups, are excluded from the performance in favour of high value others (cf. Zukin 1990, 1995).

Hemmingway (1999) insists that under the right political circumstances leisure can play a key role in enabling democratic citizenship through enhancing the 'social capital' and democratic capacities of individual citizens. This means moving beyond the instrumentality of 'representative democracy', which markets political choices to citizens via a political elite in the same way 'as advertisers attempt to persuade consumers to make certain purchases' (Hemmingway 1999, 153), to embrace 'participatory democracy', whereby citizens engage in government in their everyday lives. This would be 'a communicative, educational process in which the issues and interests confronting the community are illuminated and the abilities of citizens to participate are steadily enhanced' (Hemmingway 1999, 153–4). In the context of urban tourism and leisure, rather than their relative benefits being sold to the populace, citizens would take a proactive role in the construction of policy in the first instance, and in the distribution of subsequent benefits and costs. In Foucaudian language, this would mean replacing a disciplinary gaze through which subjects are drawn into the apparatus of government with a 'democratic gaze' that is directed by citizens who can shape and steer government from the communities and spaces within which they live their everyday lives.

This raises the issue that participatory public space is intimately linked to the creation of a healthy public realm (Sennett 1974, 1994; Jacobs 1961, 1984). However, both Sennett and Jacobs argue that in a quest to stimulate private consumption, modern urban development and spatial management have eroded participatory public spaces in cities by encouraging economic individualism at the expense of social intermingling and spontaneous contact between strangers. For Sennett (1974, 1994), truly public urban spaces are about encouraging contact between different social groups, and for Jacobs (1961, 1984), the safety, civility and vibrancy of city streets are proportional to the mix and amount of people on them. Arguably, then, rather than the development of urban tourism bracketing everyday local life from city centres, as many critics complain, the conditions for socio-spatial alienation are set before tourist enclaves take up a particular form. Furthermore, tourists could contribute to urban life more fully but for their corralling in special enclaves that value private consumption over public vitality (cf. Edensor 2000).

For Lefebvre (1996), the main issue is not about certain rights of certain people to be in certain spaces, but about rights to urban life itself. Here, play, creativity and the capacity of all urban dwellers, whether permanent or temporary, to be fully immersed in the city's rhythms of daily reproduction are what give cities vitality and meaning (Bridge and Watson 2002). In this context, space is not a container of activity in the way that mainstream planning and management view it, but a socially produced dialectical relation. The democratic challenge of Lefebvre's 'rhythmanalysis' is to reflect on the city's rhythms rather than impute spatial order and to 'connect the powers of orchestration to the rhythms of urban life and then to analyse their spatial manifestations' (Hughes 1999, 132).

Tourism, Urban Governance and Public Space in York

This section presents a concrete example of how the critical issues raised above are played out in a 'real world' setting. It analyses the way in which heritage tourism development in York city centre has proceeded in recent years to help create a central public space that is subject to increasing rounds of governance from above that meets with varied support, acceptance and resistance from below (Figure 11.1).

Figure 11.1 Performing York: Governance and Visuality in a Historic City
Source: With permission of Steve Watson

The Research Process and Methodology

All data and commentary on tourism in York cited in this chapter are in the public domain, originating from a range of sources published between 1995 and 2005. These consist of reports published by the York Tourist Bureau, York City Council, and the Joseph Rowntree Foundation, as well as seven academic papers published in international journals. The academic papers are: Madrigal (1995), Meethan (1996, 1997), Snaith and Haley (1999), Voase (1999), Augustyn and Knowles (2000), and Mordue (2005). The chapter is further underpinned by ethnographic research undertaken in York between 1996 and 2004 in two phases. Between 1996 and 1999, field observations were made during twenty separate visits to the city in order to assess how the historic core is valued and potentially contested by what Cheong and Miller (2000) call 'the tripartite of brokers, locals and tourists'. In-depth interviews were conducted in York city centre in the summer of 1996 with 35 domestic and international tourists, and another two individual in-depth interviews were carried out with residents who were mistaken for tourists. Two resident focus groups were later held with 12 people, and five 'key informant' depth interviews were conducted with 'professional brokers'. These were: the Chair of Leisure Services for York City Council, the Economic Development Officer for York City Council, the Chief Executive for the York Tourism Bureau, the Director of Attractions for the York Archaeological Trust (YAT), and a former chair of the York Group for the Promotion of Planning. All interviews were tape-recorded, transcribed and interrogated via the software package HyperResearch (see Mordue 2005 for a comprehensive analysis of this field research).

Between 1999 and 2004, nine more visits were made to York city centre in order to research issues around public space, tourist space, and the spatial politics of managing an increasingly commercialized city centre. Four of these field visits lasted for five days each, with the remaining two comprised of day visits. During this time, all the major tourist attractions in the historic core were visited annually, photographic and video records were made of tourist activity in the historic core, and spontaneous conversations with tourists and locals were annotated in field diaries. In addition, key informant meetings were held annually over the period with: the Chief Executive for the York Tourism Bureau, the Director of Attractions for YAT and the Economic Development Officer for York City Council.

All respondents who took part in this research were either identified by networking in York or were met spontaneously through participant observation. They were invited to express their views because of their willingness to participate, and because of the positionality of the information they could offer as opposed to the 'typicality' or 'representativeness' of the sample of people themselves, as is usual in quantitative research (see McCracken 1988; Geiger 1990; Cook and Crang 1995 on ethnographic methods). The interviews and discussions were underpinned by topic/issue schedules to ensure consistency, although they were sufficiently flexible and conversational to allow respondents autonomy in bringing up and opining on issues they thought important. The whole process ceased when issues and opinions became repeated to the point of saturation, meaning that continuing to uncover and explore issues and opinions further with other people could confidently be deemed as unnecessary (see Cook and Crang 1995). The respondent comments cited here are taken from Mordue (2005) for consistency of approach in relation to the way data is cited from other sources, and are presented to qualitatively illustrate bodies of local opinion that are supported by the findings of other research undertaken in York.

An Old Place under New Management

York is both a post-industrial city and an internationally recognized heritage city. It has a population of 181,094 (City of York Council 2005), which continues to expand. In terms of population profile, Falk and King suggest that 'York is surprisingly average ..., reflecting England as a whole', with a fifth being 'classified as living in poverty' (2003, 18). Until relatively recently, manufacturing was the mainstay of York's economy, with tourism being of little significance until the mid 1960s, when a tourist infrastructure began to emerge (Meethan 1996). Between 1981 and 1996, manufacturing employment dropped from 17,800 to 8,500, while tourism-related employment rose from 4,100 in 1981 to 9,279 in 2003, at which time it represented one in ten of the workforce (Census of Employment, NOMIS 1998; York Tourism 2005). These relative positions are reflected spatially in the way the city is divided into what Meethan calls 'the industrial and the pre-industrial zones', demarcated by the old city walls that envelop the historic central core (1996, 327).

Heritage attractions such as the Minster, the Jorvik Viking Centre, the Castle Museum, the medieval street layout, various guided tours, and historical events and festivals, all lie within the historic core, which now also bears many of the physical and visual enhancements associated with such places. The city draws in something like four million tourists each year – of which 19.5 per cent are from overseas, mostly from North America and Europe (York Tourism 2005). This level of activity has also meant that the centre has 'taken off as a 'leisure shopping' destination ... [with] much bigger and better shops than its catchment area could support' (Falk and King 2003, 8). Furthermore, Tempest (2005, 3) reports that 'though York's history and heritage is the largest single reason why visitors come to York', in 2004–5, 38 per cent of all direct visitor spend was on shopping, while around 10 per cent went on visiting attractions. Outside the old city walls lives more than 80 per cent of York's population (Falk and King 2003), and there exists a mixture of features like public and private housing estates, factories and business parks, offices, and 'out of town' shopping centres, familiar to the everyday environs of any post-industrial British city.

Meethan (1996) points out that since the early 1980s, York's historic core has gone through an identifiably post-modern tourism development phase, and has many characteristics typical of Edensor's (2000) 'tourist enclave'. It is a highly regulated space replete with the material and visual paraphernalia of a typical destination of its kind: tourist signage, pedestrianised streets, street entertainments, period street furniture, gentrified shops and cafes, and a host of local trails telling a themed 'spatial narrative' of multiple consumption opportunities (Zukin 1990). Such dramaturgy has been encouraged by a pragmatic city council which, in the face of a declining manufacturing base, fundamentally changed its position on tourism from one of suspicion about the 'candy floss' nature of the industry and the low quality jobs it offered, to one that accepted tourism's centrality to the city's future (Meethan 1996, 1997). In 1995, this resulted in the Labour-led administration formalizing its links with the city's private sector through the formation of the 'First Stop York Tourism Partnership', which is still in operation but under the banner of 'Visit York Tourism Partnership'. The partnership is ostensibly a tourism 'growth coalition' (Molotch 1976) that spans a host of local government departments – of which the economic development unit takes the leading role – and a range of external bodies, including the Yorkshire Tourist Board, the York Tourism Bureau, York and North Yorkshire Chamber of Commerce, York Hospitality Association, and York Hoteliers, that between them represent more than 600 private members (York Tourism 2005).

The main aim of the partnership is 'the maximization of the economic and employment advantages of tourism in York to the benefit of businesses, employees, residents and visitors', with destination marketing, product development and staff training being its core functions (York Tourism 2005). While voluntary groups are not significant in the partnership, it endeavours to reach out to the local populace via marketing and PR strategies that promote the interests of tourism as being synonymous with the interests of the city as a whole (Snaith and Haley 1999). This is done through such things as publishing the regular 'Partnership Newsletter' and 'The York Tourism Times', and organizing a 'Residents' 1st Weekend', which takes place every January when residents are given free entrance to participating heritage sites (York Tourism, 2005).

The need for such tactics indicates that tourism policy and practice can be controversial with York's residents. Indeed, Meethan tells us that as far back as the mid-1970s, 'anti-tourist sentiments within the city ran high, and there were calls for the numbers to be limited or, at least, better managed' (1996, 329). Augustyn and Knowles (2000), however, show that it is not tourism *per se* that is the political issue for residents; rather it is the 'quality' of the tourists and the lack of quality employment in the local tourism sector, and, most importantly, a growing sense of disembeddedness from the city centre that many York residents feel (Mordue 2005). The City Council has publicly admitted that there are many 'resident's moans about the city centre being turned into a 'visitors' museum'' (City of York 2002), and Falk and King (2003, 12) state that many locals are concerned about York becoming a '"twin track" city in which wealthy incomers enjoy a quality of life which is far beyond the means of most residents in the suburbs'. At the same time, they report that there is local concern that the city is starting to lose its unique identity 'as, for example, Starbucks opens up as an alternative to Betty's' (2003, 12). On this, Voase (1999) argues that too much stage-management in

York city centre has resulted in a loss of authenticity that is alienating on three interconnected levels. Firstly, it alienates locals from the city centre itself because it does not feel theirs anymore. Secondly, tourists are alienated by the staged nature of their consumption (cf. MacCannell 1999). Thirdly, and again in keeping with MacCannell's insights on 'staged authenticity' (1999, 101), both tourists and locals are alienated from each other because genuine social encounters are precluded, causing locals to retreat into the 'back-stages' of everyday local life that are unavailable to tourists.

Influencing the encounter between locals and tourists has been a major aspect of the Partnership's panoptic reach. Indeed, one of its strategic aims, as First Stop York's was, is to 'develop residents' involvement in the industry and improve their contact with tourists' so they can be 'at ease with visitors' (Tourism Strategy Group 1995, 1999; Augustyn and Knowles 2000). Its training program, internal marketing and PR campaigns have been key in delivering on this front. For instance, training has focused on scripting front-line staff's interactions with tourists because, as York's Chair of Leisure Services has stated:

> There is a recognition that part of the experience for tourists coming here is the quality of the human encounter ... If a person behind the desk is rude and grumpy and badly paid and badly trained, that is not very good for the individual's business or the business of York as a whole (in Mordue 2005, 185).

While rudeness – and spontaneity – can be scripted out of hired performances, the Partnership's power is more neutered, but subtle, when it comes to locals not working in tourism. These locals, particularly working class locals living in the periphery, have a tradition of calling for more government control of tourism (Meethan 1997; Snaith and Haley 1999). This is a direct result of the growing alienation they feel from the historic core as a space given over to, and designed for, middle-class tourists and incomers who 'are in a position both spatially and financially to take advantage of the cultural facilities and specialist shopping which now dominate the city centre' (Meethan 1997, 339). Locals' protests have also been expressed 'in a more individual, if somewhat boorish, fashion, [by] giving misleading directions and information to visitors' (Meethan 1997, 339). Little wonder, then, that since its inception as First Stop York and in its later manifestations, the Partnership has endeavoured to influence encounters between locals and tourists, and to trade local spontaneity for managed predictability and, as Voase (1999) has noted, local indifference.

This said, surveys indicate that most locals are broadly positive towards developing tourism in the city despite the problems (Madrigal 1995; Augustyn and Knowles 2000; City of York 2005; York Tourism 2005). The PR and marketing strategies have evidently been successful in getting the pro development message across, because Augustyn and Knowles (2000) remark that little of the Partnership's effort has been about the actual management of tourism impacts in the city. However, local support for tourism has always been pragmatic, qualified and ambiguous, and needs to be seen in context of the city's changing economic circumstances (Madrigal 1995; Meethan 1996, 1997). That this economic realism sits uncomfortably with local people's connectedness to the city centre is illustrated by the words of one York resident:

> I was walking through Parliament Street the other day and there was a Highland group playing bagpipes at one end and some Brazilian/Spanish samba band or something at the other end. I mean that's fine and a lot of people like it [but] what's it got to do with York? (Mordue 2005, 191).

Local feelings on these issues are anything but clear-cut or unanimous (Madrigal 1995; Mordue 2005). For instance, another resident's comments illustrate that such tourist performances can also be enjoyable local performances: 'when I'm in the city I quite like to hear all these sounds and make my way through Parliament Street, I like to see people from different countries, I like the cosmopolitan atmosphere and think York is a very parochial, dull place without such things' (Mordue 2005, 191). Yet another resident criticized such cosmopolitanism as being only aesthetic, akin to a cultural muzak that does little to penetrate the social and cultural purity of the historic core. For her, the city centre was not true to the city as a whole but a spatial and visual signifier of the people who predominantly manage, administer, control and consume it: 'It's not really cosmopolitan ... There is a massive cultural issue that the city has become to be seen as a pretty place, a middle-class, white place ... All the tourism policies are somehow culturally steered to that, all those elements are emphasized' (2005, 189).

Other residents seem much more concerned with the narrower question of who should consume the historic core rather than what is being consumed or what its staging represents. As one put it, '[t]he key issue is quality ... What we need are people who are going to stay overnight ... If you concentrated on the value end of the market, rather than the cheap day-tripping, you would also have a lot less congestion'. Supporting such sentiments, another said '[t]here just isn't enough money going around from what I call day-trippers rather than tourists, because they come here in vast numbers from the West Riding'. For these residents, rights to the city centre are all about spending capabilities rather than rights of citizenship, and given that most day-trippers will live within daily traveling to and from York, it is the most local visitors that these discourses of exclusion are trained upon. These are not uncommon opinions either (see Mordue 2005), as Meethan (1996, 329) also indicates that day-trippers have long been at the sharp end of anti-tourist sentiments in the city because they are seen by many 'to provide little to the local economy'.

First Stop York empathized with such views, even though surveys consistently show two thirds or more of all York's visitors come from the 'ABC1' socioeconomic groupings – that is, groups like managers, business proprietors and professional members of the service class (City of York 2005; York Tourism 2005). Under the heading 'Partnership Successes', it proudly reports that since the partnership started 'longer visitor stays' are up by an average of 14.8 per cent and visitor spend is up by 22.4 per cent (York Tourism 2005). It also states that the long term success of tourism in the city 'is not a matter of visitor volume: we want visitors who stay longer [and] spend more money' – making it clear that day-trippers and anyone without spending power will remain well outside the Partnership's social radar.

Craig (2003), however, questions such narrowness of purpose and doubts whether the current economic development trajectory of the city can even begin to meet the social needs of York as a whole. He points out that in 'an apparently reasonably prosperous city' it is an indictment that around a fifth of its population are living in poverty, a 'situation [that] has changed little in the past 60 years' (2003, 52); and that it is imperative for the council to address the issue of poorer residents feeling and being excluded from the city centre. He concludes that:

> [T]he costs of development associated with general economic change or specific tourist and heritage-related initiatives have to be assessed and apportioned between tourists and local residents ... at the very least a stronger focus on 'social tourism' which makes explicit the costs and benefits to local residents might help to protect poor residents from costs which are not of their making (2003, 57).

While Craig does not outline what is meant by 'social tourism', it implies something much more equitable than is happening at present or what the Partnership is planning to happen. For Madrigal (1995), the answer for York, and cities like it, is to dismantle growth coalitions like the Visit York Partnership because they only serve the narrow interests of their membership, not the community at large. And in step with much of Hemingway's (1999) views, he makes a case for the adoption of 'participatory planning', arguing that 'an informed citizenry is critically important in making decisions related to tourism development', and that '[r]ather than merely trying to convince residents that tourism is good for them, local officials should attempt to address the needs of the various constituencies existing in their community' (Madrigal 1995, 100). Alternatively, Voase (1999) champions the development of more 'non-tourist sites' in York centre, based on the type of leisure options provided by the creative and cultural industries. He argues that creative and cultural industries will satisfy the consumption demands and sociability needs of locals and tourists alike because 'the atmospheric benefits which they potentially offer ... creates a vibrancy of social relations which injects a new authenticity into city life' (1999, 295). This is the language of 'urban renaissance' (Bianchini 1990) but with a hope that cultural industries will be egalitarian in the way that their global consumer appeal cuts across already established local/tourist boundaries.

Although Craig's (2003) and Madrigal's (1995) recommendations are a long way from influencing current policy in York, recent developments indicate that the city's managers share aspects of Voase's vision. For example, First Stop York, in partnership with the City Council and the regional development agency, Yorkshire Forward, embarked on an initiative entitled 'Renaissance' which aimed 'to develop "state of the art" installations in public spaces ... to combine cultural objectives with those of the tourism and creative industries in York' (York Tourism 2005). Projects involved an evening trail for families around the city centre based on children's stories set in York, and various artistic exhibitions and performances that were illuminated at night. This desire to open up night time York spawned another essentially visual initiative called 'York:Light', which is a 10-year plan to use high-tech lighting to illuminate 'the city's major buildings and heritage sites, streets, parks, open spaces and riverside walks'. This produces a highly atmospheric visualization of York and opens up spaces, and therefore markets, that would not be accessed in the evening by consumers. The aim is to 'enhance the safety and prosperity of the city centre for residents and visitors alike' by increasing 'footfall' in the centre at night which in turn will increase 'the level of informal surveillance' (York Tourism 2005). Although there is perhaps some social value in these strategies, they are more about stretching the temporalities of city centre consumption and increasing capacity than increasing the social mix of people in it (cf. Jacobs 1961, 1984; and Sennett 1974, 1994). Indeed, these new rounds of consumption are unlikely to challenge the social purity of the city centre at all but extend the Partnership's panopticon in a way that is self-policing and economically efficient both in terms of investment and cost of operation.

Overall, however, the development trajectory of the city seems to remain indifferent to the concerns that Madrigal (1995), Voase (1999) and Craig (2003) raise. Although laudable, their recommendations are anyway overly trained on the local/tourist dualism, which sweeps past the most pernicious structural divides in York that are not so much between locals and tourists but between people of different capabilities in accessing cultural and economic power. Tourism in York has only exacerbated inequalities that have existed in the city for a long time, in that they have taken on new aesthetics and a new spatial order, particularly during its 'postmodern development phase' (Meethan 1996).

From the Specific to the General: Lessons from York

While many of the issues that York faces are particular to its situation, there are others that are 'indicative of problems facing many cities at both macro- and micro-levels' (Meethan 1997, 340). There are several interrelated aspects that are important here, which raise issues that are undeveloped in the leisure and tourism literature, and therefore warrant further discussion in this final section of the chapter. These are: the axiomatic status of the local/tourist dualism, the escalation of socio-spatial mobility, the changing nature of urban citizenship, and the governance of urban public space.

Producing neat socio-spatial categorizations, as Lefebvre (1996) contends, is symptomatic of a positivist gaze that avoids much of the complexity of human spatial relations in favour of seeking manageable spatial order. Similarly, the local/tourist dualism can be an over-simplistic imposition that has commonsense appeal but one that hardly grasps the dynamics of the social circumstances and performances through which varied people contest urban space (Chaney 2002). This type of ordering all too frequently reduces the social complexity in and around tourism development to other dualisms like the cost/benefits of so-called negative and positive tourist impacts, and so on. In these discourses, tourists tend to be configured as outsiders invading local space, upsetting some pre-existing authenticity or order therein. My own research in York, however, demonstrates that such binary divides can be more assumed than real. Moreover, local discourses and practices aimed at addressing so-called tourist impacts in York tend to distil to the social exclusion of apparently low value groups in order to accommodate higher spending consumers of leisure, tourism and culture in the way that Degen (2003) identifies. Whether York's excluded are locals or tourists, hardly seems to matter. Furthermore, perspectives that reify the local/tourist dualism as the primary source of conflict fail to acknowledge` the new (dis)order in which distinctions like 'home' and 'away' are collapsing as a consequence of the greater mobility that is sweeping the globe (Urry 2002). Tourists are only one aspect of this mobility.

The ability to be mobile as a matter of individual choice is, of course, unequally distributed. Bauman (1997) uses the metaphors of vagabonds and tourists to illustrate the inequities, where vagabonds are located on one end of the mobility continuum as enforced travellers pushed into the margins to eke out their existence, while tourists travel for pleasure as their preferences dictate. Mobility is thus dependent on one's position in the post-modern hierarchy, which is not just about the level of choice we have regarding corporeal movement, but about the ability to perform culturally and to determine one's direction and position in life and the places that we are able to inhabit and visit on life's journey. The 'perfect tourist', in Bauman's terms, is the ultimate post-modern consumer, rich both in economic and cultural capital, who represents the ideal that cities such as York now try so hard to attract as residents and tourists. Once again the issue is not about tourists' needs versus locals' needs, in the way that many commentators contend, but about rights to the city itself as Lefebvre's (1996) spatial dialectics contend.

To talk of rights to the city is to talk about citizenship, and Urry (2000) demonstrates how the private interests of consumers and the public rights and obligations of citizens have been dissolved by the flows and mobilities of nascent post-Fordism. Developments like the rise in consumer rights, the increased delivery of public services by private organizations and the way in which the state has formed governance partnerships with private, voluntary and quasi-public organizations have all impacted on what has traditionally been considered as citizenship, fixed by territorial belonging (cf. Marshall 1992). In these circumstances, Urry argues that traditional citizenship is replaced by a more hybrid 'consumer citizenship' and that the role of the state can now only regulate the performances and standards of the various organizations that deliver on this front, because its territorial boundedness gives it less jurisdictional purchase on the contemporary flows and movements of capital and people. Public spaces have thus become public/private hybrids, or 'highways of consumption' where the ably mobile dictate the performative, economic, and therefore public, terms. The less advantaged groups play more passive public roles, often as spectators of the visual and material effects of tourism, watching from the pavements and sidewalks as the world of mobility rushes by.

This is an evolving set of circumstances, and in York, for example, conflicts are evident in the way residents protest about it being a 'twin track city' and a divided city (Meethan 1996, 1997; Falk and King 2003) while others call for more market cleansing of the historic core in order to increase its social purity (Mordue 2005). The partnership, however, as might be expected, sees the private interests of its partners as synonymous with the public interests of the city as a whole. The hybridity to which Urry (2000) refers is thus not a completed state of affairs by any means but is constitutive of an ongoing struggle between varied private and public rights and interests.

Harvey (1989a,b), in common with Degen (2003), argues that in the post-Fordist city public/private coalitions have meant that the public service ethics of the old urban elite have been reduced to focusing more on urban aesthetics whereby the ideals of 'becoming' are forsaken for a state of 'being' that has no real hope or expectation of constructing a better urban future for all. For Goldberger (1996, 135), 'this produces an urbanity that is without hard edges, without a past, and without a respect for the pain and complexity of authentic urban experience. It is suburban in its values, and middle class to its core' (Atkinson 2003, 1841). The 'real' city is thus something mobile urban dwellers are generally well guarded from. Living or staying in the better parts and having good access to private transport, they can pass-by the less palatable districts and avoid the 'urban jungle' as they make their way to work, travel to and from the airport, or relax in comfortable houses and hotels (see Urry 2000 on 'automobility' and citizenship). In this context, mobility is safe and domesticated (Atkinson 2003), and city life is framed by the car windscreen or some other environmental filter.

On the other hand, Amin and Thrift (2002, 135) note how there is increasing concern about the erosion of urban public space as a backlash to 'encroaching privatization and urban dereliction', which is represented by the perils of things like gated communities and secluded zones, ghettoization, increased public surveillance, and the segregation of once communal areas. Urban leaders are therefore coming under growing pressure to recover the situation through initiatives like the reintroduction of cafes, fairs and bazaars in public spaces, ending urban dereliction, planning more multifunctional urban spaces and generally stimulating the kind of intermingling and social interaction that Sennett (1974, 1994) and Jacobs (1961, 1984) advocate. For cities like York, then, it might be possible to imagine an urban future in which the full spectrum of local people, tourists and other urban consumers are invited to share the city centre on equal terms, creating an urban cosmopolitanism that chips away at established divides.

This, however, seems a hope too far given the structural nature of the divides in York (Craig 2003) and the way in which they are reproduced performatively both in official and unofficial quarters. As Bauman provocatively deliberates:

> The postmodern setting does not so much increase the total volume of individual freedom, as redistribute it in an increasingly polarized fashion: intensifies it among the joyfully and willingly seduced, while tapering it almost beyond existence among the deprived and panoptically regulated (1997, 33–4).

The intransigence of the issues also leads Amin and Thrift (2002) to doubt whether more intermingling of strangers would be enough to produce a progressive 'politics of propinquity' (Copjec and Sorkin 1999). It may produce a much more vibrant urban aesthetics and perhaps even new public spaces of tolerance and sociability, but 'there is a limit to how far the new cosmopolitanism can be tied to the city's public spaces' (Amin and Thrift 2002, 137). The problem is that views that focus on fostering greater social interaction in public spaces imply clear separation between the public as civic sphere on the one hand, the state as institutionalized power on the other, and the market as being purely private. The real urban picture, according to Amin and Thrift, is much more fluid than this, not least because there is no single public or undifferentiated 'public good' that can be applied equally to all urban dwellers, and anyway 'what makes us sure that the piazza is the space of effective politics, so long as the real political decisions are made in the palazzo?' (2002, 136). So what might a new citizenship be built upon that could better negotiate the difficulties and opportunities of contemporary urban tourism?

Primarily, political energy needs to be directed at the structures and the socioeconomic dynamics of the new urban governance as well as on creating more democratic urban spaces of participation. Tourism is not antithetical to this and can add its own rhythms to the multiple harmonies and discords in urban life. Thus, rather than calling for better 'tourism management' strategies in urban settings, a democratic gaze should be trained upon how tourism development policy and practice sit with urban governance more generally. For this, a more participatory form of urban democracy is needed that embraces the differences and the difficulties that intersect the mobilities and the spatial fixities of urban life, rather than pursuing initiatives that regulate, sanitize and domesticate the urban experience for some while alienating others. As the York case shows, urban dwellers lacking economic and cultural capital can all too easily become disenfranchised from the city, not because of tourism, but because they do not fit the partial map of the world that coalitions like Visit York, and their supporters, insist on drawing.

References

Amin, A. and Thrift, N. (2002), *Cities: Reimagining the Urban* (Cambridge: Polity Press).

Astleithner, F. and Hamedinger, A. (2003), 'Urban Sustainability as a New Form of Governance: Obstacle and Potentials in the Case of Venice', *Innovation* 16: 1, 51–73.

Atkinson, R. (2003), 'Domestication by Cappuccino or a Revenge on Urban Space? Control and Empowerment in the Management of Urban Spaces', *Urban Studies* 40:9, 1829–43.

Augustyn, M. and Knowles, T. (2000), 'Performance of Tourism Partnerships: A Focus on York', *Tourism Management* 21, 341–51.

Bagnall, G. (2003), 'Performance and Performativity at Heritage Sites', *Museum and Society* 1:2, 87–103.

Bauman, Z. (1997), *Postmodernity and its Discontents* (Cambridge: Polity Press).

Bayliss, D. (2004), 'Ireland's Creative Development: Local Authority Strategies for Culture-Led Development', *Regional Studies* 38:7, 817–31.

Bianchini, F. (1990), 'Urban Renaissance? The Arts and the Urban Regeneration Process', in B. Pimlott and S. MacGregor (eds), *Tackling the Inner Cities: The 1980s Reviewed, Prospects for the 1990s*, 215–50. (Oxford: Clarendon Press).

Bianchini, F. (1991), 'Re-Imagining the City', in J. Corner and S. Harvey (eds), *Enterprise and Heritage: Crosscurrents of National Culture*, 212–34. (Routledge: London).

Bridge, G. and Watson, S. (2000), *Companion to the City* (Oxford: Blackwell).

Bridge, G. and Watson, S. (2002), *The City Reader* (Oxford: Blackwell).

Census of Employment, *NOMIS* (1998).

Chaney, D. (2002), 'The Power of Metaphors in Tourism Theory', in S. Coleman and M. Crang (eds), *Tourism: Between Place and Performance*, 193–206. (New York: Berghahn).

Cheong, S. and Miller, M. (2000), 'Power and Tourism: A Foucauldian Observation', *Annals of Tourism Research* 27, 371–90.

City of York (2002), A Hotel Tourist Guide to York [Online]. Available at: http://www.cityofyork.com [accessed: 18 November 2002].

City of York (2005), A Hotel Tourist Guide to York [Online]. Available at: http://www.cityofyork.com [accessed: 6 July 2005].

City of York Council (2005), The City of York Homepage [Online]. Available at: http://www.york.gov.uk [accessed: 6 July 2005].

Coleman, S. and Crang, M. (eds) (2002), *Tourism: Between Place and Performance* (New York: Berghahn).

Cook, I. and Crang, M. (1995), *Doing Ethnographies* (Durham: CATMOG).

Copjec, J. and Sorkin, M. (1999), *Giving Ground: The Politics of Propinquity* (London: Verso).

Corner, J. and Harvey, S. (1991), 'Introduction: Great Britain Limited', in J. Corner and S. Harvey (eds), *Enterprise and Heritage: Crosscurrents of National Culture*, 5–20. (London: Routledge).

Craig, G. (2003), 'Managing a Historic City in the Interests of its Residents: An Anti-Poverty Strategy for York City Council', in N. Faulk, and F. King, *A New Vision for York*, 48–57 (York: Joseph Rowntree Foundation).

Degen, M. (2003), 'Fighting for the Global Catwalk: Formalizing Public Life in Castlefield (Manchester) and Diluting Public Life in Al Raval (Barcelona)', *International Journal of Urban and Regional Research* 27:4, 867–80.

Duncan, J.S. (1999), 'Elite Landscapes as Cultural (Re)productions: The Case of Shaughnessy Heights', in K. Anderson and F. Gale (eds), *Cultural Geographies*, 53–69. (London: Longman).

Duncan, N.G. and Duncan, J. (1997), 'Deep Suburban Irony: The Perils of Democracy in Winchester County, New York', in R. Silverstone (ed.), *Visions of Suburbia*, 161–79 (London: Routledge).

Edensor, T. (2000), 'Staging Tourism: Tourists as Performers', *Annals of Tourism Research* 27, 322–44.

Edensor, T. (2001), 'Performing Tourism, Staging Tourism: (Re)producing Tourist Space and Practice', *Tourist Studies* 6:1, 59–81.

Falk, N. and King, F. (2003), *A New Vision for York* (York: Joseph Rowntree Foundation).

Foucault, M. (1991), 'Panopticism', in P. Rainbow (ed.), *The Foucault Reader*, 206–13. (London: Penguin Books).

Foucault, M. (1997), *Discipline and Punish: The Birth of the Prison* (New York: Vintage Books) (trans. by A. Sheridan).

Geiger, S. (1990), 'What's So Feminist About Women's Oral History?', *Journal of Women's History* 2:1, 169–82.

Goldberger, P. (1996), 'The Rise of the Private City', in J. Vitullo Martin (ed.), *Breaking Away: The Future of Cities*, 101–38. (New York: The Twentieth Century Fund).

Griffiths, R. (1993), 'The Politics of Cultural Policy in Urban Regeneration Strategies', *Policy and Politics* 21:1, 39–46.

Halewood, C. and Hannam, K. (2001), 'Viking Heritage Tourism: Authenticity and Commodification', *Annals of Tourism Research* 28, 565–80.

Hall, C.M. (1994), *Tourism and Politics: Policy, Power and Place* (Chichester: Wiley).

Hannigan, J. (1998), *Fantasy City: Pleasure and Profit in the Postmodern Metropolis* (London: Routledge).

Harvey, D. (1989a), 'From Managerialism to Entrepreneurialism: The Transformation in Urban Governance in Late Capitalism', *Geografiska Annaler* 71, B:1, 3–17.

Harvey, D. (1989b), *The Condition of Postmodernity: An Enquiry into the Origins of Cultural Change* (Oxford: Blackwell).

Healey, S., Cameron, S., Davoudi, S., Graham, S. and Madani-Pour A. (eds) (1995), *Managing Cities: The New Urban Context* (Chichester: Wiley).

Hemingway, J.L. (1999), 'Leisure, Social Capital, and Democratic Citizenship', *Journal of Leisure Research* 31:2, 4–27.

Hughes, G. (1999), 'Urban Revitalization: The Use of Festive Time Strategies', *Leisure Studies* 18, 119–35.

Jacobs, J. (1961, 1984), *The Death and Life of Great American Cities: The Failure of Town Planning* (London: Peregrine Books in association with Jonathon Cape).
Lefebvre, H. (1996), *Writings on Cities* (Oxford: Blackwell).
Lipietz, A. (1993), *Toward a New Economic Order* (Cambridge: Polity Press).
Logan, J. and Molotch, H. (1987), *Urban Fortunes: The Political Economy of Place* (Los Angeles: University of California Press).
MacCannell, D. (1976), *The Tourist: A New Theory of the Leisure Class* (Berkeley: University of California Press).
Madrigal, R. (1995), 'Residents' Perceptions and the Role of Government', *Annals of Tourism Research* 22:1, 86–102.
Marshall, T.H. (1992), 'Citizenship and Social Class', in T.H. Marshall and T. Bottomore, *Citizenship and Social Class*, 17–27. (London: Pluto Press).
McCracken, G. (1988), *The Long Interview* (London: Sage).
Meethan, K. (1996), 'Consuming (in) the Civilized City', *Annals of Tourism Research* 23, 322–40.
Meethan, K. (1997), 'York: Managing the Tourist City', *Cities* 14:6, 333–42.
Meethan, K. (2001), *Tourism in Global Society: Place, Culture, Consumption* (Basingstoke: Palgrave).
Molotch, H. (1976), 'The City as a Growth Machine: Toward a Political Economy of Place', *American Journal of Sociology* 82, 309–22.
Mordue, T. (2005), 'Tourism, Performance and Social Exclusion in "Olde York"', *Annals of Tourism Research* 32:1, 179–98.
Robins, K. (1991), 'Tradition and Translation: National Culture in its Global Context', in J. Corner and S. Harvey (eds), *Enterprise and Heritage: Crosscurrents of National Culture*, 21–44. (London: Routledge).
Sennett, R. (1974, 1994 2nd edn), *The Fall of Public Man* (London: Faber).
Shaw, G. and Williams, A. (2004), *Tourism and Tourism Spaces* (London: Sage).
Smith, L. (2006), *Uses of Heritage* (London, Routledge).
Snaith, T. and Haley, A. (1999), 'Residents' Opinions of Tourism Development in the Historic City of York, England', *Tourism Management* 20, 595–603.
Stoker, G. (2000), Introduction, in G. Stoker (ed.), *The New Politics of British Local Governance*, 1–10. (London: Macmillan).
Tempest, I. (2005), *Tourism Trends* (a report presented to York City Council's Economic Development Board 27th September 2005).
Thrift, N. (2004), 'Driving in the City', *Theory, Culture and Society* 21:4/5, 41–59.
Tourism Strategy Group (1995), *York Tourism Strategy and Action Plan* (York: First Stop York).
Tourism Strategy Group (1999), *York Tourism Strategy: Three Year Action Plan 1999–2002*, (York: First Stop York).
Tulloch, J. (1999), *Performing Culture* (London: Sage Publications).
Tunbridge, J.E. and Ashworth, G.J. (2000), *Dissonant Heritage: The Management of the Past as a Resource in Conflict* (Chichester: Wiley).

Urry, J. (1995), *Consuming Places* (London: Routledge).

Urry, J. (2000), *Sociology Beyond Societies: Mobilities for the Twenty-First Century* (London: Routledge).

Urry, J. (2002), *The Tourist Gaze: Leisure and Travel in Contemporary Societies* (London: Sage Publications).

Voase, R. (1999), 'Consuming' Tourist Sites/Sights: A Note on York', *Leisure Studies* 18, 289–96.

York Tourism (2005), First Stop York [Online]. Available at: http://www.york-tourism.co.uk [accessed: 8 July 2005].

Zukin, S. (1990), 'Socio-Spatial Prototypes of a New Organization of Consumption: The Role of Real Cultural Capital', *Sociology* 24, 37–56.

Zukin, S. (1995), *Landscapes of Power: From Detroit to Disney World* (Los Angeles: University of California Press).

Chapter 12
The Tourist as Juggler in a Hall of Mirrors: Looking Through Images at the Self

Tom Selwyn

One aim of this chapter is to explore the idea that significant strands of both tourism marketing and tourism itself are underpinned by a dialogue between image-makers and tourists on the nature of the self. This aim is implicitly underwritten by familiar assumptions that tourists and tourism operate within philosophical frameworks made up of relations between, *inter alia*, self, persons, places, natural and cultural landscapes, objects, products and the multifarious images of all of these. A second aim, developed towards the conclusion of the chapter, is to suggest that the conversations about the self articulated by the kinds of images to be examined here are embedded, with surprising precision, within the ideological shape and structure of the global politico-economic environment in which contemporary tourism takes place. The overall purpose of what follows is thus to indicate a starting point for wider discussions on the dynamics in the tourism world of the relations between the personal and emotional/cognitive on the one hand, and the global and the political/economic on the other. In various ways the chapter pursues lines of work consisting of anthropological/semiotic readings of promotional and other forms of tourism related imagery (see for example Dann 1989, 1996; Selwyn 1993, 1996; Edwards 1996; Pritchard and Morgan 2003; Pink 2007).

The chapter involves looking at a selection of images. The majority of these are taken from tourist brochures subtracted from the 2004 and 2008 World Travel Marts[1] in London augmented with two collages of recently available postcards from the cities of Ferrara and London. Readers may (with some justification) question the validity of this selection procedure: the limitations are self-evident but difficult to avoid. In defence of the method (if such a term is not too pompous) we may recall that the outstanding exemplar of the genre of scholarly readings of contemporary advertising consisted of a collection of images carefully chosen by the author himself (Berger 1983) to tell, and then illustrate, his own story. This is more or less the approach adopted here.

1 Three images are taken from brochures dated before 2004.

Echo, Narcissus and the Tourist as Philosopher

From the early days of tourism studies, social scientists have decorated 'The Tourist' with a variety of epithets: MacCannell's (1976) neo-Durkheimian, Graburn's (1977) 'pilgrim', Cohen's (1979) 'drifter', Gottleib's (1982) 'king', Dann's (1989) 'child', and so on. The suggestion made here is that, aided and abetted by specialists in imagery whose employment hinges on knowing intimately both about internal personal and external landscapes (and thus how to attract visitors to particular locations), tourists are also philosophers. This unoriginal thought merely extends to the field of tourism Charles Taylor's (1989, 21) observation that there is no one in the world who operates outside a philosophically defined framework although, as he adds, not everybody is either able or willing to articulate this framework in the same way as a philosophically-trained observer might. The tourist, the generic and undifferentiated category encompassing all types of the species, is no exception to this rule. The additional suggestion made here, however, is that the philosophical field of enquiry in which our tourist is most likely to be encountered is that which starts from a concern with the construction of the self.

The suggestion is thus that in the tourism sphere, as in others, we find advertising specialists and agencies, together with other impresarios of the tourist imagination, grounding much of their image design work on the fact that there is literally no one (thus no tourist) who is unconcerned, consciously and unconsciously and on a continuing and permanent basis, with the nature of his/her self. But such an interest is socially and culturally mediated: an interest in the self by, for example, British or Northern European tourists is moulded by a cultural disposition and habitus, to use terms made familiar by Pierre Bourdieu, steeped in British and/or Northern European experience (but see Pink 2007, 35 for a discussion on theories of photography that bears directly and critically on these sort of assumptions). In the final part of the chapter, I will indicate how such mediation, which is political and economic as well as cultural, might be described in the present case.

However, we know from the myth of Echo and Narcissus, as well as Plato's allegory of the cave, that we come to know ourselves partly from images reflected on surfaces (from lake and/or wall in these two cases). Perhaps, therefore, the most appropriate metaphor for the world inhabited by our philosopher-tourist is a space hung with mirrors that reflect outlines and aspects of the self. Viewed in this way – and whether we encounter him or her in the bars and beaches of Magaluf (Andrews 2000, 2009) in Italian galleries or museums (Clark 2004, 2005), or in the fields, canyons and rivers of the adventure tourist (Selwyn 1996) we may see our tourist as a juggler in a hall of mirrors.

The Self and a Grammar of Tourist Imagery

The senses we have of our selves emerge from the moment of birth, in the process of relations with significant others and the landscapes we inhabit. Throughout life we move through spaces filled with people, objects and images that provide the means constantly to form, re-form, frame, and re-frame senses of our selves and understandings of who we are and/or who we would like to be. The tourism field, especially the field of tourism imagery, is one that incorporates many such people, objects and landscapes and it is quite plausible to suggest that we could in time describe the rules and structures of a much larger and more complex grammar governing our imaginations as we journey through this field. The completion of such a large project is, however, for another day. At this point we have the limited objective of suggesting how an interest with the self lies at the outset of such a possible future endeavour. It is high time, therefore, to consider the notion of self.[2]

Notwithstanding Dann's[3] beguiling observation that it would be difficult for anyone to post a letter who did not know what a mailbox was, with its even more seductive implication that we must therefore be prepared to describe the essences of things, no attempt is proposed here to come up here with a definition of the self. In fact we might begin with the observation that the essence of the idea of self is, precisely, its elusiveness. What we do need, however, is a sketch indicating some of the routes of an enquiry,[4] one way of setting up which is to locate the notion of self within a field of other terms that, whilst in no way synonymous with it, have direct and indirect relationships to the idea. This holds out the possibility of looking at the intersections and juxtapositions between these other terms as they overlap, combine and re-combine, in the process of leading us closer to the self.

From a wide range of possible terms, we may pick on three, each associated with a companion term that help us establish our framework. These are: (a) body, together with individual; (b) collectivity, with its companion identity; and (c) exchange, accompanied by hospitality. We may start with the body. Bodies are not selves. Nevertheless, especially at a time when biometric identity cards are in the making, with information about a person's DNA or iris structure inscribed on, or in, them, the relation of the physical body to the social self is clear enough. In the tourist context, of course, bodies play symbolically prolific roles. The term individual, accompanied as it normally is by a sense of uniqueness, is a term we may readily see as a close companion to both body and self. Indeed, in a contemporary Anglo-American/Northern European context it sometimes seems that the self consists of the embodied individual imaginatively sculpted by the rhetoric of such iconic terms as freedom, democracy, choice and (of course) the market itself. Such a view is challenged, however, by the continuing presence, wherever we look, of images, signs and symbols reminding us of the persistence, even in a disenchanted post-modern world, of senses of belonging and collective identity. Selves need

2 What follows borrows from the section (Selwyn 2005) in the MED-VOICES website <http://www. med-voices.org> on the idea of person.

3 Made during the presentation of a conference paper (Dann 2004) on what Aristotle might have taught Alexander the Great.

4 The main inspiration for this part of the chapter is Taylor's *Sources of the Self.*

groups to be themselves. This leads directly to the third term, exchange. Selves, needing recognition, come into being in the course of relationships with others. In the tourism sphere, one of the closest companions to the idea of exchange is hospitality, for it is precisely within the laws and conventions of hospitality, the exchange of goods and services between hosts and guests, that self and other is conceived, embodied and practised.

The purpose of these brief reflections is to suggest that to talk of the self involves entering a terminological framework in which body, collectivity and exchange – each with their companions of individual, identity and hospitality – play determining roles. In the sphere of tourism imagery, the proposition is thus that a sense of self emerges out of the textures of such imagery as these are shaped by these terms and the ideas and values that they contain and connote.

Images of Self within a Triad of Body, Collectivity and Exchange

Having prepared the ground we may now look at a selection of 18 images, arguing that we (and/or, more importantly, the tourist reader/observer) may find an outline of the self within a triad of three clusters depicting the body (with individual), the collectivity (together with identity) and exchange (and hospitality).

Bodies and Individuals

Figure 12.1 (image 1) consists of images featuring bodies from two postcards sold in the centre of the tourist circuit of the Italian city of Ferrara. The juxtaposition of the playful semi-clothed cyclists of one postcard and the clothed medieval-looking person in another makes the girls very slightly subversive in the context of a historic town most famous for being, and primarily marketed as, a site of medieval pageants and an annual *pallio* second in importance only to Siena's. We will come back to them later.

Figure 12.1 (image 2), taken from a brochure advertising Zanzibar, places the individual at the centre of the world. It self-consciously plays with the idea of *Sultan*. Here, it is the name of a five-star hotel, but we know from another time that this was an appellation attached to a ruler to whom all were hierarchically subordinated. Within this tourist seascape, however, the single body (significantly of a woman) renders the historical connotations with the Sultan even more striking – for *this* individual is clearly mistress of her own destiny. No *seraglio* for her. The image thus speaks with an ironic (dis)regard for the history of the area to an omnipotent individualism that celebrates the triumph of *homo aequalis* (in a Dumontian sense) and world system within a setting in which assuredly has no truck with supreme rulers.

Figure 12.1 (image 3) depicts a robed couple on the balcony of a four-star hotel promoted by Thomas Cook. The picture belongs to a familiar and extensive genre of tourist images in which white and healthy heterosexual couples occupy centre stage. Just visible on the horizon (distant and indistinct) are the roof tops of the

Figure 12.1 Image 1; Image 2; Image 3 and Image 4

village houses, but our couple is more-or-less screened from that view by palm fronds that effectively isolate their balcony from any built surroundings. They appear alone in a comfortable jungle. The portrait is made up of intimations of feelings that speak of the proximity of the personal and intimate.

The personal experience offered by Thomas Cook finds another expression in Figure 12.1 (image 4). This promises a 'very personal experience' of the city of Lisbon. The reason that this abstract image fits within a basket of images having to do with body and individual stems precisely, if paradoxically, from the fact that there are no people (nor any nature or culture come to that) shown. It is thus entirely open to the reader/viewer's own interpretation and imagination and is made intelligible only by, to repeat the phrase, the 'very personal experience' that can reveal anything and everything, including questions about whether the design is a flag, whether it is under water, and what, if anything, is connoted by the squares.

In many respects, though not quite all, the grapes and texts of Figure 12.2 (image 5) place even more emphasis on the primacy of the individual body than the design of Figure 12.1 (image 4).

Figure 12.2 Image 5

Here the promise is of an engagement by 'all the senses' – the sight, hearing, touch, taste, smell of the tourist's eyes, ears, fingers, tongue, and nose – as he/she approaches the *terroir* of Rioja. Post Durkheimian, post pilgrim, drifter, king, or child, the potential tourists to Rioja are offered a total sensory engagement with the seductive grapes of the image by regional authorities announcing their wish for nothing less than the entire body of their visitors. The one socially and economically anchoring feature of the image, however, stems from the self-evident fact that grape production involves careful organization and orientation to the cultural and natural contexts of the territory in which they are grown. In this sense, therefore, whilst the bunch of grapes appears to invoke the primacy of individual physical sensations, there is also an un-stated implication in the image that such sensations sit within a social context.

By the time we reach the heroic figure of Figure 12.3 (image 6), we have travelled very far from the playful *biciclettes* of Ferrara, although we have reached what some visitors, and potential visitors, to Blaenavon, South Wales, might well declare was a cathartic moment in the display of the body as an attraction and symbolic vehicle for tourist experience. If the postcards of Ferrara imply a gently challenging freedom of the body from cultural convention, Blaenavon responds with a powerful statement that the individual body reaches heroic potential primarily when placed in the forefront of productive service of the wider community. In this pastiche of Soviet realism, our miner is working for the good of the whole, not primarily for himself. The World Heritage status of the site effectively bestows (or perhaps more accurately gives back) the value and honour to coal mining and coal miners that was taken from them by the combined forces of mine closures, the battles between unions and government in the 1980s, and the general sentiment that global warming is linked to the burning of fossil fuel. Figures 12.1–3 are all variations on the linked themes of body and individual, whereas Figures 12.4–6 play on themes of collective identity.

Collectivities and Collective Identities

Figure 12.4 (image 7) is a painting by Chagall to which the Israeli brochure designers (rather than Chagall)[5] have attached the label 'Jerusalem'. The painting depicts joyful aspects of Jewish life in Eastern Europe. With its trees, flowers and birds, it is predominantly agricultural in tone. Klezmer musicians mingle with dancers and young couples. The Israeli flag appears, whilst the overall youthfulness and *joi de vivre* of the painting itself is emphasized, in the photograph shown in the brochure, by the young European-looking couple in front of it. The painting clearly speaks of collective identity – or, to be more accurate, three overlapping identities: Jewish Eastern Europe, the state of Israel and the city of Jerusalem. Pre-

5 It is worth stressing that Chagall himself is a painter whose work has contributed more than most to Jewish/Christian understanding. In this sense he is a cosmopolitan painter very far away from the narrower 'nationalist' visual agenda suggested by the brochure in question.

Figure 12.3 Image 6

figuring the discussion below on the politico-economic context, we may notice only that this mythic East European cultural landscape is far away from the deeply divided city that contemporary Jerusalem actually is today.

Figure 12.4 (image 8), from a Romanian tourist brochure, testifies to the presence in tourist brochures of images of collective life. Romanian tourist authorities are not alone in thinking that images of collectivities – here of families and villages

Figure 12.4 Image 7; Image 8

– are attractive to potential tourists. The Arabian forts of Figure 12.5 (image 9), taken from a brochure distributed by the tourist offices of Oman, evoke imagined traces of fiercely independent tribal and/or ethnically defined groups. Again, lightly to pre-figure later discussions, we may refer briefly to other brochures from the Gulf, in which the advantages are extolled of holding conferences in this modern, peaceful part of the Arabian peninsula. The good governance, safe and peaceful investment climate promoted by these brochures are not in any way challenged by the forts in this picture. Precisely the reverse is the case. Resonating with references to the desert warfare of tribal pasts, the forts present the business traveller with a tamed taste of orientalist exotica: just right for the conference's weekend 'cultural programme'.

If Figure 12.4 (image 7) speaks of bucolic country life in Eastern Europe, Figure 12.4 (image 8) of familial rites of passage, and Figure 12.5 (image 9) of tribal groupings, Figures 12.5 (image 10)–Figure 12.6 (images 11 and 12) play imaginatively with a variety of collective identities that include global cultural cosmopolitanism (the Mona Lisa with an Indian *tika*), a British colonial world with an inbuilt potential for its own reversal (the British flag attached to a tea bag in an Arabian tea cup) and versions of full frontal British nationalism. Thus, Figure 12.5 (image 10) ironically punctures the idea of the Mona Lisa as an exclusively or pre-eminently Italian and/or European icon by decorating her forehead with the mark of a Indian married woman, whilst Figure 12.6 (image 11) inverts the West's

Figure 12.5 Image 9; Image 10

Figure 12.6 Image 11; Image 12

(including Britain's) dependence on Arabian oil by appearing at first sight to evoke British colonial power (Arabia marked by the British flag) but doing so with a subliminal mockery that, on second sight, might suggest that this time around it is Britain herself in the Arabian cup: is Britain itself the tea bag?

But Figure 12.6 (image 12), a collage of popular images from postcards sold in London, returns British nationalist power to unequivocal glory. Swimmers wear costumes of national flags whilst their images on postcards are routinely placed in the same stand as postcards of the Queen, Princess Diana, red-coated guardsmen and Buckingham Palace. All of these images are well and truly 'nationalized'. And if anyone were to doubt the nationalist significance of the ubiquitous 'Full English' (as the eggs, beans and pork sausages and bacon are known), they need look no further than Andrews' (2000) comprehensive analysis of the political and symbolic importance of the cooked breakfast in the British tourist malls of Mallorcan charter tourism resorts.

The few paragraphs above have sought to explore and illustrate the assertion made earlier that selves need collective as well as individual identities The collectivities in question here have included: rural community, family, village, tribal *ethnie*, continent (as in the case of the 'European' Mona Lisa), colony, nation, and state. We may now move to the third of our clusters.

Exchange and Hospitality

The central question that the images of this cluster raise concerns the terms of the relationship between tourist and the landscapes and people that he/she visits. How, and with what meaning, in other words, is the exchange between imagined host and guest articulated?

Figure 12.7 (image 13) depicts a Croatian farmers' market. The image evokes an atmosphere for the potential tourist nostalgically to recall benign exchange relationships in which produce is locally grown, and trade conducted by people who belong to the same community. The brochure reader/viewer – the tourist or potential tourist – is offered grapes by a stallholder. Tourists appear (if they appear at all) merged with local shoppers. All seem part of the same social and cultural system.

Figures 12.7 (image 14) and 12.8 (image 15) develop the theme of exchange relationships based on degrees of sociability and hospitality (the latter announced textually by the Ethiopian brochure in the first of this series) and gift giving. In all these cases, however, the theme of mutuality of exchange – an indivisible part of hospitality, classically conceived (cf. Pitt-Rivers 1977, 94–112) – is accompanied by an undercurrent of awareness of the unequal power and wealth of host and guest. The Ethiopian hostess (Figure 12.7 [image 14]) offers tea with herbs, the Gambian couple (Figure 12.8 [image 15]) a drink in a tin mug. But the implication must be that the tourist and/or potential tourist brochure reader/viewer has a great deal more than either of these items. In these senses the images turn out to be fundamentally ambivalent.

Figure 12.7 Image 13; Image 14

The exchange relationships between host and guest in these images speak, at one level, of mutuality and hospitality. Figures 12.8 (image 16) and 12.8 (image 17), however, assert or reassert (as many tourist brochures do) the primacy in tourist exchange of the commercial market. In an Air New Zealand aeroplane (Figure 12.8 [image 16]), we are offered Pacific Islands to consume as food, whilst in Hong Kong (Figure 12.8 [image 17]), the injunction to the young(ish) white couple is FIND IT. The implication here is that more or less anything and everything may be both found and bought. For some, perhaps a majority, of tourists from the Western world, tourism involves, or itself becomes, a form of shopping. Finally, Figure 12.8 (image 18) completes the sheaf of images in our collection. Here is a white man, at ease with himself and his can of Zambezi beer, posing in front of what we may presume to be the Victoria Falls. The image resonates with generations of earlier pictures of white men posing in front of natural wonders, including animals killed in hunting. It resonates too with images, in paintings and photographs, of land and property owners.

The Juggler in the Hall of Mirrors

Looking at, and interpreting, brochures and postcard images in the ways I have begun to do above gives an indication of the depth of ideas on offer to tourist-

Figure 12.8 Image 15; Image 16; Image 17 and Image 18

philosophers about the nature of the self. To start with, there are the many masks and characters that can be chosen by the viewer/observer to experiment with in contemplating the nature of his or her self in relation to others. These range from the 'pure' individual of the Lisbon senses and the Zanzibar beach, to one that is part of Thomas Cook's version of a couple, to that of an imagined partner in the mutual exchanges of the Croatian farmers' market or the hospitality practices of

Ethiopia and Gambia as portrayed by their tourist authorities. Then there is the question of the relation of the self to social and collective structures. Designers of the 'Jerusalem' or Romanian images, for example, seem to be aiming to draw the reader/observer into distinct cultural or community groups that are there, so to speak, for the joining. But the Hong Kong and, above all (with considerable brilliance), the Pacific Island authorities look towards consuming selves who are either or both masters and mistresses of the tourist market for all types of goods, on the one hand, or hugely powerful consumers (who can eat whole islands at one go) on the other.

This leads to the dazzling array of collectivities to which our tourist may imagine him/her self to be attached: the nation states of the London, Arabia, and Israeli postcards and brochures; the tribal *ethnies* of Oman and Arabia (and there are many examples of such brochures produced by African tourist offices), or the families and village communities of Romania. There are even faint echoes of the British and/or other empires. But beneath the surfaces on which the above images appear there is a second layer of issues that further stretch the imagination of our tourist/brochure reader. Amongst these are two that merit particular attention, namely the tourist's disposition implied by the images towards time and history, on the one hand, and place and space on the other.

The Ferrara images, for example, raise questions about the relation of the present to the past – a topic that is, of course, a staple of tourism and 'heritage' studies. What, for example, is our relation to medieval Ferrara? Is that medieval person an ancestor of ours? Do we belong to the same lineage as he does? If so, in what sense do we? If not, why do we not? In what sense are his values our values? Is our identity closer to those of the bold girls? At the other end of the spectrum of images of the body stands the Blaenavon miner. In front of this heroic man, with the collection of ideas and values associated with him, the visitor is placed on the spot, as it were, facing issues at once more immediate, more remote and definitely more painful than those raised by the knight from Ferrara. After all *our* society, culture, nation, state and empire were all predicated, to a significant extent, on coal, coal mines and miners, and the mining communities in the Welsh valleys. And exactly where are these now? Amongst the unemployed and dispossessed? Forgotten relics of an industrial history that fuelled prosperity until the arrival of the 'service economy'? So we may ask a similar question of the miners as we asked of the medieval figures of Ferrara: Do *they*, the beaten but here heroic miners, belong to the same lineage as us? If they do, though, how did we keep faith with them when the combined forces of technological and economic change and a vindictive government were arraigned against them? Some might say we abandoned them until it was safe to build them a World Heritage Site. And as for the Arabian forts, do not they appear to us rational, scientific, Western, and democratic citizens of nation states as primitive reminders of former social and cultural arrangements that we have long ago graduated from?

As for place and space, most of the images provoke questions about our relation to the world and the land and landscapes that are depicted in the brochures.

Our swimmer walking on the Zanzibar beach may seem to own the coast and 'Zimbabwe No. 1' (of all places) more certainly seems to own the Victoria Falls, but the designers of the British and Israeli brochures leave potential tourists with little doubt that the British monarchy (here standing for the British state) and the Israeli state own London and Jerusalem.

Furthermore, beside questions of ownership and control, there are some more subtle issues having to do with space that the images evoke in the minds of the brochure reader/viewer. One of the more arresting of these has to do with the relation between the public and private spaces depicted in the imagery. For example, would the Ferrara girls be wholly acceptable in an actual street in the city (rather than on a postcard) for example? If less acceptable, where does this leave the tourist? After all we have a mass of ethnographic evidence that tourists and tourism are routinely held by residents of tourism destinations to be responsible for re-drawing boundaries between public and private – sometimes to the perceived detriment of the society in question.

Here, then, is an indication of the richness of the field of images at the service of our tourist as we find him or her in a hall of mirrors, using depictions of a world that seem to offer multiple choices about the role of his or her self in the scheme of things. As a tourist/potential tourist it seems that I can elect to place myself almost anywhere within the widest imaginable spectrum of social formations and structures – from village families to desert tribes to medieval city guilds. In this sense, our brochures and postcards seem almost to take on characters of philosophical treatises offering us complex arrays of elementary questions about our identity and identification and position in time and space. However, moving towards the conclusion of the chapter, we come to yet another layer of interpretation that enables us to reveal that the dominant narrative woven by our images has a more contemporary edge to it, that the search for self is conducted within a framework shaped by political and economic features of the modern world system.

Of Contemporary Nation States, Individuals and Markets

Moving towards a conclusion, we may now attempt a more rounded and generalized view of the field. The argument here is that, despite interesting and important exceptions to the rule (which we will come to first), the majority of our images, when considered together, tell a story of the political dominance of nation states (some of which contain ambivalent traces of former and contemporary empires) and the centrality of individual consumers in a global market system. Starting, then, with the exceptions to the general rule, we may now make a tour amongst the sites and landscapes our images promote. We will move amongst our images in a different order from the one taken earlier, and may start and finish with a single individual body.

Our starting point this time around is the Welsh village of Blaenavon marked by the striking image of a coal miner (Figure 12.3). As already suggested, this heroic (for Wales, talismanic) worker carries connotations, some of which have been set out above. His image is part of a permanent exhibit in the Blaenavon World Heritage Site. The town, site of the 'Big Pit' coal mine, is now a museum that in 2009 mounted an extensive exhibition on the 1984/5 miners' strike. There are other exhibitions and archaeological reminders in the town of the famous iron works (opened in 1788, closed in 1900) and steel works that produced railway lines for networks around the world. Our miner is a fine example not only of the service that miners gave their communities and the wider economy but also of the British industrial revolution itself. In the terms of the present analysis, we thus have an individual body that embodies and celebrates the role in the British industrial revolution of the working class and the class and communal solidarities associated with it.

Community solidarity is also the theme of the depictions of rites of passage in our Romanian village (Figure 12.4 [image 8]) and, arguably, the Croatian market (Figure 12.7 [image 13]). Individuals in both of these appear as integral parts of wider collective structures and processes and the ideas and values that these carry. We are clearly in sites that *could* be pre-modern as they invite us to look back to a historical period when villages, families, and production were linked in a territorial and immediately visible way.

We might note at this point that these particular images in no way imply that all Croatian or Romanian tourist images are collectivist in tone. Whilst, as former socialist states (or parts of such) and therefore historically familiar with collectivist imagery, both countries, as modern nation states with the open economies that EU membership demand, are also definitively part of the modern world system. Flavours of a nationalist kind, moreover, are not unfamiliar to Croatian tourism imagery. But we are concerned here only with our present collection of images. These, the argument runs, take us (and the tourist/viewer) along trajectories that range from the collectivist to the definitively individualist.

The world of our Mona Lisa (Figure 12.5 [image 10]) is one in which (as in this case) a European painting can become, with a minimum of fuss, an Indian one. In this way the image self-consciously disrupts the connection between art and either private or national ownership, often made in promotional literature from national tourism offices. Instead it opens up the possibility in the tourist's imagination of an open cosmopolitan universe in which artistic and other borders may easily be crossed. The proposition is thus that these four images, in slightly different ways, speak to a relatively open and collectivist world. All of them draw social spaces within which individuals may find a place and over which they do not dominate. The collectivities at issue comprise class (Blaenavon), family (Romania) and village/small town (Croatia). In these three, as in the Mona Lisa, which arguably evokes a global collectivity, we are far from the emphasis on nation, consumption, and the market to which we will come now.

As already noted, the Ethiopian (Figure 12.7 [image 14]) and Gambian (Figure 12.8 [image 15]) images foreground hospitality. The names of the two states feature prominently, effectively 'nationalizing' the hospitality in the process. As already noted above, however, questions remain both about the nature of hospitality in a context of a tourist market in which there are pronounced inequalities of wealth and power between host and guest (is 'real' hospitality actually possible in such an economic climate?), but also about the role of states in managing this market. In the Gambian case the presence of the middle-aged woman reflects the fact that one well known market niche in Gambian tourism is one in which single (and richer) white women tourists seek relationships with younger (and poorer) Gambian men (Yamba and Wagner 1986). Given the textual prominence of the name Gambia in the image, one is left wondering about the role of the state in this relationship. Once we get to the Omani forts (Figure 12.9) and the Arabian cup with the (British) 'nationalized tea bag' (Figure 12.6 [image 11]), the relative poverty of host and guest has been clearly reversed, but we remain with a nationalist idiom. Indeed, if we were to look at a much broader selection of tourist brochures from the Gulf states, including Oman, Dubai and UAE, we would find that, in announcing that this is a region of states that have built high quality shopping experiences (or words to that effect) they present a heady brew of 'nationalized marketing'. As such, they reflect the role of the region's states as the West's valued allies in the protection and promotion of a world built on consumption – with tribal forts and identifications being firmly put in their place (as amusements to engage with in between proper shopping, business meetings, and/or corporate team building).

There are few images that convey the importance in the tourism business of consumption with such force and dark humour as that by Air New Zealand (Figure 12.8 [image 16]). The advertisement is, of course, a joke from many points of view, not least that the Pacific Islands are by no means as powerless as the image suggests.[6] But, as O'Rourke (1987) has shown, the make-believe world of Western tourists in or over Pacific Islands is routinely shaped by ambivalent cannibalistic fantasies. The brilliance of the Air New Zealand brochure designers is to have come up with a message to the effect that *they*, the Pacific Islanders, might well have been cannibals once but in today's market-led world it is we Western travellers who hold both the power and inclination to eat them.

Whilst the above five images introduce, directly and/or by implication, the tourist to the global market place in which they are invited to roam, the Jerusalem (Figure 12.4 [image 7]) and London (Figure 12.6 [image 12]) images confirm the leading roles of states (here Israel and the United Kingdom) in this world. In these two cases, as in many others in the genre of images we are considering, it appears that the state itself is the offer. Our collection of images thus presents us with narrative of a world in which (at least) food, culture, history, hospitality,

6 The largest single grouping in the United Nations General Assembly is that made up by island states.

and persons appear as objects for consumption. It is precisely at this point that we come, finally, to the individual and the individual body.

Rioja (Figure 12.2) introduces our final set of images by making the bold statement that all the senses of the individual body lie at the heart of tourism whilst Lisbon (Figure 12.4) follows this lead by asserting that visiting the city involves very personal experiences. Thomas Cook's couple (Figure 12.1 [image 3]) and the Hong Kong shoppers (Figure 12.8 [image 17]) develop these themes, the former by stressing the capacity of the hotel to provide a space for sensuous intimacy, the latter by combining intimacy with shopping. Ferrara (1) juxtaposes the unbound bodies of the contemporary bicycling girls with the clothed historical figure, suggesting in the process the idea of contemporary freedom in the face of dimly remembered social and cultural imperatives that have successfully been consigned to annual processions and/or 'heritage sites'.

We come, finally, to Zanzibar (Figure 12.1 [image 2]) and Zimbabwe (Figure 12.8 [image 18]). The former combines many of the elements previously identified. Our heroine clearly evokes a sense of individualism, sensuality, and freedom. And, as suggested previously, her presence on the beach seems proprietorial, a quality about which there can surely be no doubt in the case of 'Zimbabwe No. 1'. This young man is arguably the trump card in our collection which now, taken as a whole, appears as a focussed summary of the modern world system of liberal, democratic, states underpinning a global regime of free markets in which we, as tourists, may roam as free consumers. At ease with himself and his beer, our hero offers the comforting thought that *we* really can own, control, and consume the world and its natural treasures and resources.

Conclusion

This chapter began by stating the case for seeing the tourist as one who is interested, in a philosophical way, with the nature of him/her self. The argument was advanced that tourists use imagery from brochures, postcards, and (by implication) elsewhere to disentangle and then relate various aspects of self. It was suggested that the key opening terms of what might be termed a 'grammar' of engagement of self with the world included body, individual, collectivity, identity, exchange, and hospitality. Travel and tourism are frequently associated with 'getting away' from the everyday in order to encounter the possibilities to experiment with one's identity and sense of self. At first sight our own collection, in which we find pictures of medieval knights, expansive coastlines, tropical vegetation, representations of desert tribes, islands, waterfalls, and other natural and cultural sites of wonder and distinction, suggests that the metaphorical field for philosophical reflection and experiment is almost limitless. Yet, on second viewing, our images appear more sober and restrictive. Rather that being vehicles for journeys of untrammelled imagination they reveal themselves as guardians of a particular narrative about the order of the world. Thus, out of 18 images, we find four that self-consciously integrate self and collectivity, past and present, cultural difference and continuity. In a variety of ways, the remaining 14 consist of landscapes (highly imaginative, to be sure) in which the tourist/viewer finds his or her scope to play and experiment with the self circumscribed by confining boundaries that are drawn by the insistent primacy of the individual, state, and market.

References

Andrews, H. (2000), 'Consuming Hospitality on Holiday', in C. Lashley and A. Morrison, (eds) *In Search of Hospitality: Theoretical Perspectives and Debates*, 235–54. (Oxford: Butterworth Heinemann).

Andrews, H. (2009), 'Tits Out for the Boys and No Back Chat: The Gendering of Tourist Space in Magaluf', *Culture and Space*, 12:2, 166–82.

Berger, J. (1983), *Ways of Seeing*, (London: BBC and Penguin).

Clark, D. (2004), 'Jewish Museums: From Jewish Icons to Jewish Narratives', *European Judaism: A Journal for the New Europe*, 3:2, 4–17.

Clark, D. (2005), 'The Field as 'Habitus': Reflections on Inner and Outer Dialogue', *Anthropology Matters (e-journal).*

Cohen, E. (1979), 'A Phenomenology of Tourist Experiences', *Sociology*, 13, 179–201.

Dann, G.M.S. (1989), 'The Tourist as Child: Some Reflections', *Les Cahiers du Tourisme*, (Aix-en-Provence: Centre des Hautes Etudes Touristiques).

Dann, G.M.S. (1996), 'The People of Tourist Brochures', in T. Selwyn (ed.), *The Tourist Image*, 61–82. (Chichester: John Wiley).

Edwards, E. (1996), 'Postcards: Greetings from Another World', in T. Selwyn, (ed.), *The Tourist Image*, 197–222. (Chichester: John Wiley).

Gottlieb, A. (1982), 'Americans' Vacations', *Annals of Tourism Research*, 9, 165–87.

Graburn, N. (1977), 'Tourism: The Sacred Journey', in V. Smith, (ed.) *Hosts and Guests: The Anthropology of Tourism*, (Philadelphia: University of Pennsylvania Press).

MacCannell, D. (1976), *The Tourist: A New Theory of the Leisure Class*, (Berkeley and Los Angeles: University of California Press).

MED-VOICES (2004), 'The Person', [Online]. Available at http://www.med-voices.org [accessed: 1 July 2009].

O'Rourke, D. (1987), *Cannibal Tours* (film) (Canberra: O'Rourke and Associates).

Pink, S. (2007), *Doing Visual Ethnography*, (London: Sage).

Pitt-Rivers, J. (1977), *The Fate of Shechem, or the Politics of Sex*, 94–112. (Cambridge: University Press).

Pritchard, A. and Morgan N. (2003), 'Mythic Geographies of Representation and Identity: Contemporary Postcards of Wales', *Tourism and Cultural Change*, 1:2, 111–30.

Selwyn, T. (1993), 'Peter Pan in South-East Asia: Views from the Brochures', in M. Hitchcock, V. King, and M. Parnwell, (eds) *Tourism in South-East Asia*, 117–37. (London: Routledge).

Selwyn,T. (1996), 'Atmospheric Notes from the Fields: Reflections on Myth Collecting Tours', in T. Selwyn, (ed.) *The Tourist Image*, 147–61. (Chichester: John Wiley).

Taylor, C. (1989), *Sources of The Self*, (Cambridge: University Press).

Yamba, C.B. and Wagner U. (1986), 'Going North and Getting Attached: The Case of the Gambians', *Ethnos*, 51:3–4, 199–222.

PART IV
Constructing Place

Chapter 13

The Story behind the Picture: Preferences for the Visual Display at Heritage Sites

Yaniv Poria

Research on heritage tourism sites often focuses on the display or, in other words, the patent as opposed to latent heritage. This line of research generally highlights the concept of 'power relations', emphasizing the stakeholders' impact on the presentation and interpretation of certain events presented at heritage tourist attractions. Not surprisingly, such studies repeatedly find that the winner's version of history is on display, while the loser's account has essentially vanished (or has been archived far from the public eye). Such findings are not surprising given that studies of the history of history reveal an identical pattern: the winners retain the power to authorize their version of history. This chapter adopts Tunbridge and Ashworth's approach, which asserts that the study of heritage settings '... must shift from the uses of heritage to the users themselves and thus from the producers (whether cultural institutions, governments or enterprises) to the consumers' (1996, 69). Specifically, the chapter highlights the visitor's expectations of the visual display rather than providing an analysis of the visual display. No attempt will be made to analyse how heritage site management (which in some cases is comprised of a single individual) decides about the identity and manner of the visual (picture/photograph) display, but rather, the chapter explores what the visitor would like (or not like) to see and why. The conceptualization of people's preferences of visual displays is based on studies centring on people's experiences of diverse heritage sites conveying diverse meanings to visitors. Examples of such sites include the Wailing Wall, Massada, Jerusalem, The Anne Frank House and Auschwitz Concentration Camp (some of which have been the focus of tourism literature studies: see, for example, Poria, Butler and Airey 2003, 2004a,b; Poria, Reichel and Biran 2006, 2007, 2009). In addition to being identified with a particular heritage, many of the aforementioned sites, as with other heritage attractions, are considered 'must-see' or 'must experience' tourist attractions or are located on the way to, and from, other tourist attractions. As will be discussed later in the chapter, these aspects are important for understanding people's experiences at heritage sites and their expectations of the interpretation of those sites.

This chapter asserts that research on the experience of spaces presenting cultural historic artefacts should be based on the relationships between a site's attributes and visitors' perceptions of the heritage and its presentation. It will be suggested that only some of those spaces should be captured as heritage spaces, and some heritage spaces may not include cultural historic artefacts. Additionally, this chapter directly addresses a topic often latent in the literature; namely that heritage sites and the visitors to such sites often aim to differentiate and create social borders among peoples. In other words, heritage and heritage tourism is often a resource for furthering conflict and not for conflict resolution. This chapter regards visits to heritage sites as the epitome of the *heritagization process*, a social process whose final outcome is the presentation and interpretation of heritage (rather than the archiving or sustaining of heritage). *Heritagization* is a process in which heritage is used as a resource to achieve certain social goals. As approached here, heritagization is not about the past but about the use (and abuse) of the past to educate – and at times inculcate – the public (this approach differs from the common tourism literature approach, see Inglis and Holmes 2003). Additionally, this chapter does not overlook the fact that heritage tourist attractions are frequently visited for reasons other than interest in the represented heritage. For example, such sites do not charge admission fees, are located proximal to other *lighter* types of tourist attractions, and, similar to McDonald's, the toilets can be used for free. A further reason for visiting heritage tourism attractions, commonly ignored in the tourism literature, is the mitigation of shameful perceptions associated with hedonistic tourist activities. Also, many tourists consider the visitation of heritage sites as part of their social duties and thus, such sites are often classified as *must-see* or *must experience* tourist attractions.

In pointing to the diverse roles of the visual display at heritage settings, this chapter aims to provide a better understanding of people's experiences of heritage settings by clarifying elements such as visitor perceptions of the site, visitation motives, expectations from the visit and the perceived benefits of the visit. The chapter challenges several working assumptions common in associated literature, as well as the actual practice of interpretation at heritage sites. It seeks to contribute to the theoretical understanding of heritage tourism by challenging the view that only a naïve search for nostalgia or a simplified commoditized past(s) are at the core of heritage tourism. The chapter also aims to highlight the role of heritage sites in the heritagization process and indicates the importance of the emotional involvement during the heritage experience, in contrast to the commonly discussed cognitive or entertainment experience. To address such expectations towards visual displays, the following components will be discussed: (1) the definition of a heritage site; (2) the role of interpretation; and (3) the role of the visual display, while distinguishing between visual display and textual interpretation. The supporting data is based on literature that centres on heritage (cf. Lowenthal 1998) heritagization (cf. Smith 2006) and the politics of heritage (cf. Harrison and Hitchcock 2005; Moore and Whelan 2007), in addition to literature focusing on heritage sites management (cf. Timothy and Boyd 2003; Leask and Fyall 2006), museum management (cf. Fopp 1997; Hooper-Greenhill 2000; Serrell 2006) and interpretation (cf. Beck and Cable 2002). Additionally, data about the conceptualization of the visit is based on studies centring on people's experiences of settings that are classified as heritage tourist attractions and those that visitors perceive as heritage sites. These sites are visited by a diversity of tourists, some of whom regard the site as part of their own heritage, some who regard the site as part of someone else's heritage, and others who may not be familiar with the site's heritage at all.

Is a Heritage Site a Historic Site?

Contrary to what is commonly asserted in heritage literature, this chapter does *not* begin from the proposition that heritage sites should be viewed as historical sites, ancient sites or those that are recognized or classified as 'heritage' sites by particular institutions or organizations. Specifically, this study does not follow the common descriptive approach to heritage tourism, but rather adopts the experientially-based approach (Timothy and Boyd 2003). Moreover, it is argued – unapologetically and in line with the experientially-based approach – that some World Heritage Sites should not be perceived as heritage spaces; indeed, some such sites although designated as World Heritage Sites, are not perceived as part of someone's heritage, not to mention the world's heritage. A heritage site, as conceptualized here, is not even required to be authentic (and, in this context, authentic is equivalent to real), even if archaeologists have determined, 'objectively' and based on scientific epistemology, that the site is really ancient. Thus, it is argued that not every historic site is a heritage site and vice versa, although it may be.

To understand the role of the visual display, a heritage site is clarified here as a site in which people are involved in a heritage experience. Such a site may be authentic, may be old, may be listed as a heritage site (even by UNESCO, the world's self-proclaimed representatives), but it need not be any of the above. Instead, and in line with the experientially-based approach to heritage tourism, a heritage site is viewed as a site that people sense and regard (rather than think) as linked to their own personal heritage. Moreover, a heritage site – as opposed to a historic site – is one that people feel belongs more to their heritage than to other people's heritage (after all, if a site belongs to a particular heritage, it is, by inference, not part of another's heritage, after Tunbridge and Ashworth 1996). Based on the approach adopted here, Ground Zero could comprise a heritage site despite the fact that it cannot be considered ancient by archaeologists or is not listed by any particular institution as such. In this case, the emphasis is on people's experience and the meanings assigned to the visit, rather than the site's attributes, per se. An outcome of such an understanding is that one site visited by different groups of people will render different meanings for different people. Some may regard a site as a heritage site, while others will not. In keeping with the above, a heritage site can also be a football field for die hard fans of a certain team. The visitations to such sites have much in common to 'real' heritage sites.

Heritage sites comprise a component of the heritagization process (Smith 2006). Heritagization, as Smith attests, is a process in which individuals establish a sense of identity (actively or passively) based on their links to a certain community or group of people. The heritagization process dictates the importance of the *affect* on the visitor and, as such, the *experience* at heritage sites, in lieu of the belief that the site's physical attributes must be perceived as authentic. Thus, a heritage site, in contrast to a historic site, can be a replica and, by definition, not the 'real thing'. For example, a site in which the stories and pictures of people's lives in concentration camps is portrayed, even if that site's venue is not in Poland or Germany, may still be a heritage site for some of its visitors, regardless of the discrepancy between the location of the actual historical events and the site's geographical venue. The important point necessary for understanding the expectations of the visual is that a heritage site is perceived by the visitors as such, rather than by the curator (this point does not exclude the curator's role in affecting the individual's perception of the site heritage).

Interpretation: Different Reasons for not Telling the Truth (or for Outright Deceit)

The visual is at the core of the tourist experience (Small 2008). The visual plays an important role in interpretation in general and in heritage settings (at the site or outside the site) in particular. To understand the role of the visual in the process of interpretation, interpretation must first be defined. A review of the literature suggests that interpretation can be defined as the transmission of information from the presenter (that is, the supplier) to the viewer (that is, the visitor) in an attempt to educate the visitor. Howard (2003, 244), for example, argues that interpretation 'covers the various means of communicating heritage to people'. Anderson and Low's (1976) definition is also of importance for this study, as it emphasizes the role of management in the provision of interpretation and the heritagization process. They defined interpretation as a 'planned effort to create for the visitor an understanding of the history and significance of events, people, and objects' (Anderson and Low 1976, 3). Note, here, the use of the words 'planned' and 'create', both of which are suggestive that while there are different heritages 'out there' in the public arena, it is up to management to decide what to present, what to archive and what to demolish.

Studies dealing with presentation at heritage sites, and the interpretation provided, commonly highlight that the heritage presented is not 'real' or 'the truth', denying the possibility that objective reality exists. The differences between the 'real thing' and the presentation are the result of two main factors. The most commonly mentioned factor is that the display aims to justify and validate a specific version of history (while ignoring other versions). Those studies, commonly initiated from a feminist research perspective, support Orwell's (1953) often quoted argument that those who control the present control the past and those who control the past control the future. The second factor, not commonly highlighted (possibly as marketing and managerial issues are not prestigious enough to be examined by 'real' scholars), refers to managerial considerations arising from the need to attract visitors. Heritage settings, due to increasing competition with other tourist attractions (including other tourist attractions classified as heritage) and the need to be profitable, have to make the 'show' more interesting and appealing. This phenomenon results in modifications and manipulations, both of which distance the presentation from the 'real thing' if that 'real thing' is not attractive enough. This competition between sites and tourists attractions leads the interpretation to be a tool to attract visitors to the site. Based on the aforementioned, the working definition for interpretation adopted here is *the process of the transmission of knowledge, its diffusion and understanding by the individual.*

This definition implies that interpretation is a process that begins with the information chosen to be presented or hidden (whether it is 'real', 'objective' or fake), and continues through the visitor's understanding and experience of the interpretation. The working definition presented thus recognizes that visitors play an active role in the interpretation process, and this recognition is important for understanding the visual's function during the visit. Additionally, it should be noted that the visual, like interpretation, has to present the winners' truth and, in certain cases, be unique in order to attract visitors.

Although not the focus of this chapter, from a management perspective the visual display has several interlinked aims. The first such aim is educating visitors of the need for protection, aimed at facilitating understanding of the importance of protection and conservation (Timothy and Boyd 2003). This line of thought is reflected in Tilden's (1977, 38) statement, 'through interpretation, understanding; through understanding, appreciation; through appreciation, protection'. Infrequently, the objective of entertaining the visitor is also mentioned. However, the entertaining experience is perceived as a means of endorsing learning (Timothy and Boyd 2003). Additionally, and in line with literature that elaborates on the links between power and the heritage presented (or hidden), interpretation almost always aims to sustain the power of the dominant hegemonic groups in society. Finally, heritage may also be seen as a means to attract visitors simply to generate revenues. As will be presented in the next section, while some of these supplier expectations reflect the visitor interest, some do not.

The Visitors: Who Wants to Learn, About Whom and About What?

Visitors to heritage cultural sites have been grouped and distinguished in various ways. Chen (1998), for example, distinguishes between visitors according to two main motives: pursuit of knowledge and personal benefit (such as relaxation, sightseeing, recreation). McCain and Ray (2003) differentiate between two sub-segments of heritage tourists: legacy tourists and other special interest tourists. Moscardo (1996) argues that the degree of one's willingness to be educated or entertained at heritage settings can be used to distinguish visitor types. Stewart et al. (1998) identify four types of visitors, partially based on the duration of interpretation sought: 'seekers,' 'stumblers,' 'shadowers' and 'shunners.' Espelt and Benito (2006) revealed differences in the interest in culture acquisition and interpretation among four groups of visitors to a heritage city: non-cultural tourists, ritual tourists, interested tourists and erudite tourists. McKercher and du Cros (2002) differentiated between visitors to cultural attractions, based on the experience sought (shallow vs. deep) and the importance of cultural tourism in the decision to visit a destination (low vs. high). The aforementioned classifications follow the descriptive approach and attempts to differentiate between visitors to cultural heritage sites. The classification reported here is based on the assumption that heritage rather than culture may be, for some, at the core of the visit. The classification suggested is based on the individual's interest in heritage rather than interest in cultural capital. In accordance with this line of thought, visitors are grouped into three categories: 'identity builders', the 'multicultural minded audience' and the 'ticking and guilt reducers'.

From the beginnings of tourism research, links between travel and identity have been revealed and rehearsed. Aitchison and Reeves (1998, 51) maintain that '... identity, meaning and behaviour are constructed, negotiated and renegotiated' while discussing the links between spaces and tourism. The argument made by Aitcheson and Reeves is useful for the differentiation of visitors based on their perception of the site relative to their own heritage. The *identity builders* do not only regard the site as part of their own heritage, but perceive the site as more strongly linked to them than to others. They are characterized by their interest in constructing an identity or strengthening an existing one. This group of tourists is busy with 'identity work' (Griswold 2004) and therefore seeks exposure mainly to its own heritage. The *identity builders* may regard other site visitors as guests who come to see and learn about the *identity builders* and their heritage. They may regard the space as holy, although they may classify themselves as secular. The interpretation should convey to them, in line with the heritagization process, who they are. More specifically, the experience should inform them who is part of their 'tribe' (a task often achieved by stating who is not part of their tribe), why and when they became and evolved from individuals into a tribe (when the tribe was evolved and for what reason). Additionally, the heritage presented should inform them, and others, in a reality in which people are becoming increasingly similar one to one another, about their present uniqueness based on elements of the past. Moreover, the interpretation should highlight why such uniqueness cannot fade away and, more importantly, why it is worth suffering (and even fighting) for. The interpretation should support their system of attitudes and beliefs, including elements such as social hierarchy and certain traditions that the individual cannot justify in ways other than by asserting 'it was always done this way'.

The *identity builders* may perceive any form of formal interpretation as a barrier to the experience, which prevents them from feeling connected to the site. Those visitors, although frequently classified as tourists in accordance with formal definitions (World Tourism Organization [WTO] definition), do not regard themselves as tourists as they do not reside in the area where the heritage site is located. Tourism is often thought of as an experience in which individuals travel *out there* (far away from their normal place of residence) to be with *others* (who are not locals) which, in turn, offers an opportunity to play a different role than in day-to-day life (as there are fewer social norms and more anonymity). The *identity builders*, in contrast, through temporary mobility learn about themselves in places rich in social norms they follow, often with people with whom they are familiar and associated with. In addition, and in contrast to what is commonly noted in the literature, *identity builders* are not seeking to enrich their general knowledge during the visit. From their perspective, knowledge may be acquired prior to the visit (at home, reading a book) or after the visit (watching a documentary). For example, Jewish Israeli visitors to the Western Wall do not perceive the visit as a tourist experience. Moreover, they feel 'at home' at the site, feeling that this space belongs to them more than to others (Poria et al. 2003). They feel that this is a place in which their tribe and their ideological framework evolved. These visitors do not come to learn about the site; rather, they come to validate the ideological framework in which they believe. For them, the visit is a 'must experience' activity (rather than a 'must see' site).

The second group of visitors to heritage sites, the *multicultural minded audience*, recognizes that a heritage site is linked with a particular heritage that is not their own. This group is interested in an enriching educational experience that does not centre on them but results in increasing their cultural capital. It is thus a group less interested in an emotional experience, representing a very different kind of visitor than the *identity builders*. While *identity builders* expect to feel certain emotions (such as happiness, sadness or empathy) as well as feelings that are associated with your own tribe (such as pride, a sense of uniqueness), the *multicultural minded audience* is not interested in the latter. This group is interested in negotiating and even challenging the meanings they give to the heritage displayed (as it is someone else's heritage and not their own). In the context of interpretation of heritage spaces, this group can be seen as a *multicultural minded audience* (Goldstein 2003, 152) that is curious about other cultures, religions and ethnic groups. These tourists are interested in learning about many heritages if those heritages can be seen at one site (and often heritage sites are multi-heritage sites). These visitors are searching for the objective truth, assuming the roles of leisure time archaeologists and historians. They are not concerned with strengthening their identity during their visit, although may reflect on it. Moreover, they may regard the *identity builders* as fanatics or fetishists, people attaching feeling to objects. For them, the site is just another place where they can observe unique social behaviours and enrich their knowledge.

The final group of tourists described here is composed of two subgroups – *the ticking* and *guilt reducers* – which, at certain sites, comprise the vast majority of tourists. Certain heritage sites are well-known tourist attractions. They are visited by *the tickers*, those who come to the site only because it is a famous 'must-see' tourist attraction (rather than a 'must experience' tourist attraction) – the fact that the site happens to be a heritage site is merely coincidental. Although almost no attention is given to this issue, many heritage sites are visited simply because they exist and admission is gratis. For this group of visitors, the site *may* be perceived as old and historic. However, their visit is a culmination of their perceived duties as tourists and may occur despite their absolute lack of interest. After all, when visiting China or Israel you have to see the Wall (the Great Wall, the Western Wall), and in Paris the Eiffel Tower is a must. Such visits are part of their duties as tourists. The *guilt reducers* are those visitors who visit sites to reduce feelings of guilt, which often are linked to hedonistic tourist activities (eating well, shopping, having sex). Both sub-groups of visitors are primarily interested in seeing the site or watching the people who regard this visit to be part of their own heritage. These tourists have lower expectations with respect to the learning process. However, they have some expectations of learning and being entertained as well (unless it is a site at which they are aware that they are supposed to feel sad or empathize with someone). These visitors will hope that the visit to the site will be short and not boring (rather than interesting).

The Role of the Visual

It must be recognized that while there are many types of sites presenting historic artefacts, this chapter discusses the role of the visual in the context of a typical museum presenting a specific group's history, and often located in a city which is a centre for other additional tourist attractions linked with the group or nation concerned (for example, the British Museum, London, The Egyptian Museum, Cairo). Nevertheless, the role of the visual is relevant to other tourist attractions, and in line with the descriptive approach, the heritage site attributes have to be taken into account when attempting to reveal the visitor-expected experience. At the Auschwitz Concentration Camp, for example, people do not expect to be entertained, as virtually all visitors recognize the site as a symbol of human evil.

The Aim of the Visual

The role of the visual is highlighted in light of the above visitor classifications as well as with respect to its contribution in comparison to other forms of display, such as text and voice interpretation. The visual aims at facilitating the visitor experience. As such, for every group the visual has a specific role. For the *identity builders*, the visual aims, first and foremost, at providing an emotional experience, endowing them with the feeling that their own, and not someone else's, heritage is on display. They yearn to feel that the displayed heritage belongs to them more than to others. Moreover, as the displayed heritage is their own, they expect the visual to provide them with a rationale for their current system of beliefs and their tribal uniqueness. This segment assigns high importance to the visual display as compared to the other two segments. The *identity builders* may even argue that they are mainly, or exclusively, interested in the visual. As suggested previously, their main motivation for visiting is not knowledge acquisition. Nor do they want to read or listen to the 'history' of the site, activities they commonly perform prior to the actual visit. The *identity builders* want to feel that they are connected to the site and that the site is connected to them more than to others. Based on studies reported earlier (Poria et al. 2003, 2006, 2007, 2009), those visitors may even regard other methods of interpretation as a barrier to their personal heritage experience. They come to experience the site and the presented artefacts as confirmations to their belief in their group's eternal uniqueness. They expect the visual to provide them with evidence and support of their tribe's common past and the superiority of their myths. The visual should support their collective memory and validate their belief in their myths, rather than assert a scientific truth (unless the scientific truth supports their agenda).

The second group, the *multicultural minded audience*, come mainly to learn about others and gain cultural capital, an aim which may be achieved through a visual that will inform and educate them. This group expects the visual to accompany and assist other forms of interpretation, such as text. The visual's aim is to facilitate the learning process and it should be used to save time and effort. If a picture can 'paint' a thousand words then, the *multicultural minded audience* would argue, we need to read fewer words. The visual should also provide evidence that the presented text is, indeed, the 'truth' and should make the educational element of the visit more entertaining. The visual should also explain happenings or assertions that cannot be explained using text alone. For the third group, *the ticking* and *guilt reducers,* visual imagery is about making the visit entertaining and, if properly executed, will cause them to remember what the site is all about. For this segment, the visual aims at reducing the risk of boredom, as well.

What Should be Visualized?

The *identity builders* would like to see people and artefacts linked to their group, nation, tribe or other affinity. Moreover, they would also like to see symbols and signs that will declare to them and others that it is their group being visualized. They would like to see elements that visualize their myths as such myths are important for legitimizing their current social order. Moreover, if their group is presently involved in actions that may lead to feelings of shame, the visual should justify the actions that cause such feelings. The visual should not highlight only the past, but also how the past is connected to the present. The visitors want the visual to provide evidence that they, the visitors, are linked and associated with those who appear in the visuals. As such, they want to see themselves, or something that represents and symbolizes them, as part of the visual. For them, the visual display does not have to be a real, authentic element, but something that causes them to feel that the visit experience is authentic, emphasizing their undoubted uniqueness as a social tribe in comparison to other groups, nations or tribes.

The second group, the *multicultural minded audience,* visits heritage sites to learn about others. They would like to see the 'real thing' as well as visual imagery that will educate them about others. This does not mean that they do not wish emotional involvement; however, the pictures must present real objective truth. The *ticking* and *guilt reducers* are not highly concerned with what is presented, unless it is interesting and unique. The visitor groups may also be distinguished by whether they expect the visual to be the truth and the 'category' of history they would like to observe. There are several ways to illustrate the relationships between history and heritage. History can be divided into four categories: *Good Active/Bad Active/Good Passive/Bad Passive. Active* history is something that I or my tribe has done. *Passive* history is something that happened to me or my tribe. *Good* is something linked to feelings most people enjoy having (such as pride), and *Bad* is something linked to feelings we do not like to have (such as shame). Those who regard the site as part of their own heritage, the *identity builders*, would like the visual not to present the *Bad Active* elements in their history. We do not like to see and visit places that demonstrate that we have done something bad of which we are ashamed. The two other groups, who do not perceive the site as part of their own heritage, would like to see the four different histories described above. The differences will be that the *multicultural minded audience* will prefer to be exposed to visual evidence which was recognized as truth, while the *tickers* and the *guilt reducers* would like to see the interesting and entertaining elements. According to the former two sub-groups, if the historical truth is not of interest and unique, there is no need to show it.

Summary and Conclusions: Explaining the Present, Clarifying the Past, or Enjoying the Moment

Expectations regarding the visual display should be captured in line with the expectations of the interpretation of the heritage in general. The *identity builders,* those visiting the site to strengthen their identity, consider the visual as a vehicle enabling an emotional experience which legitimizes their uniqueness, relative to others. The visual should differentiate the *identity builders* from others and at times even cause them to feel a sense of superiority. The visual display should create social borders, by informing viewers who and what is part of them and what belongs to others. The group who came to learn, *multicultural minded audience,* expects the display to facilitate the knowledge gathering process. It should contribute to knowledge acquisition and provide evidence that the presented narrative and events actually happened. The display should explain the past and the visual is evidence of past happenings. For the *tickers and guilt reducers,* the visual should aim to make the visit more interesting and enjoyable; after all, it is easier to see a movie than read a book.

Acknowledgments

The chapter is based on a series of studies focusing on the individual experience of heritage sites, in which I have been involved along with Professor Richard Butler, Professor David Airey, Professor Arie Reichel and Dr Avital Biran.

References

Aitchison, C. and Reeves, C. (1998), 'Gendered (Bed) Spaces: The Culture and Commerce of Women only Tourism', in C. Aitchison and F. Jordan (eds) *Gender, Space and Identity, Leisure, Culture and Commerce*, 47–68. (Brighton: Brighton, Leisure Studies Association).

Andreson, W.T. and Low, S.P. (1976), *Interpretation of Historic Sites.* (Nashville: Altamira).

Beck, L. and Cable, T. (2002), *Interpretation for the 21st Century.* (Champaign: Sagamore).

Chen, J.S. (1998), 'Travel Motivation of Heritage Tourists', *Tourism Analysis* 2, 213–15.

Espelt, N.G. and J.A.D. Benito (2006), '"Visitors" Behaviour in Heritage Cities: The Case of Girona', *Journal of Travel Research* 44, 442–8.

Fopp, M.A. (1997), *Managing Museums and Galleries.* (London: Routledge).

Goldstein, K.R. (2003), 'On Display: The Politics of Museums in Israel Society', unpublished PhD thesis, University of Chicago.

Griswold, W. (2004), *Culture and Societies in a Changing World.* (Thousand Oaks, CA: Pine Forge Press).

Harrison, D. and Hitchcock, M. (2005), *The Politics of World Heritage: Negotiating Tourism and Conservation.* (Clevedon: Channel View Publication).

Hooper-Greenhill, E. (2000), *Museums and the Interpretation of Visual Culture.* (London: Routledge).

Howard, P. (2003), *Heritage, Management, Interpretation, Identity.* (London: Continuum).

Inglis, D. and Holmes, M. (2003), 'Highland and Other Haunts: Ghosts in Scottish Tourism', *Annals of Tourism Research*, 30, 50–63.

Leask, A. and Fyall, A. (2006), *Managing World Heritage Sites.* (Amsterdam: Elsevier).

Lowenthal, D. (1998), *The Heritage Crusade and the Spoils of History.* (Cambridge: Cambridge University Press).

McCain, G. and Ray, N.M. (2003), 'Legacy Tourism: The Search for Personal Meaning in Heritage Travel', *Tourism Management* 24, 713–7.

McKercher, B. and du Cros, H. (2002), *Cultural Tourism: The Partnership between Tourism and Cultural Heritage Management.* (New York: The Haworth Hospitality Press).

Moore, N. and Whelan, Y. (eds) (2007), *Heritage, Memory and the Politics of Identity: New Perspectives on the Cultural Landscape.* (Aldershot: Ashgate).

Moscardo, G. (1996), 'Mindful Visitors: Heritage and Tourism', *Annals of Tourism Research* 23, 376–97.

Orwell, G. (1953), *1984.* (Hammondsworth: Penguin Group).

Poria, Y., Butler, R. and Airey, D. (2003), 'The Core of Heritage Tourism: Distinguishing Heritage Tourists from Tourists in Heritage Places', *Annals of Tourism Research* 30, 238–54.

Poria, Y., Butler, R. and Airey, D. (2004a), 'Links Between Tourists, Heritage and Reasons for Visiting Heritage Sites', *Journal of Travel Research* 43, 19–28.

Poria, Y., Butler, R. and Airey, D. (2004b), 'The Meaning of Heritage Sites for Tourists: The Case of Massada', *Tourism Analysis* 9, 15–22.

Poria, Y., Biran, A. and Reichel, A. (2006), 'Heritage Site Management: Motivations and Expectations', *Annals of Tourism Research* 33, 162–78.

Poria, Y., Biran, A. and Reichel, A. (2007), 'Different Jerusalems for Different Tourists: Capital Cities – The Management of Multi-Heritage Site Cities', *Journal of Travel and Tourism Marketing* 22, 121–38.

Poria, Y., Reichel, A. and Biran, A. (2009), 'Visitors Preferences for Interpretation at Heritage Sites', *Journal of Travel Research*, 48:1, 78–91.

Serrell, B. (2006), *Judging Exhibitions: A Framework for Assessing Excellence.* (Walnut Creek: Left Coast Press).

Small, J. (2008), 'The Absence of Childhood in Tourism Studies', *Annals of Tourism Research* 35, 772–89.

Smith, L. (2006), *Uses of Heritage.* (London: Routledge).

Stewart, E.J., Hayward, B.M., Devlin, P.J. and Kirby, V.G. (1998), 'The "Place" of Interpretation: A New Approach to the Evaluation of Interpretation', *Tourism Management* 19, 257–66.

Tilden, F. (1977), *Interpreting Our Heritage.* (Chapter Hill: University of North Carolina Press).

Timothy, D.J. and Boyd, S.W. (2003), *Heritage Tourism.* (Harlow: Prentice Hall).

Tunbridge, J. and Ashworth, G.J. (1996), *Dissonant Heritage: The Management of the Past as a Resource in Conflict.* (Chichester: Wiley).

Chapter 14
Site Seeing: Street Walking Through a Low-Visibility Landscape

Tim Copeland

The use of visual sources that survive as archaeological imagery in the landscape and have been disinterred from it by human or, more rarely, natural agencies is often dealt with in a simplistic way in models of constructing the past (Copeland 2004). Perhaps this is because in a period of insecurity among archaeologists about the communication of their results to non-archaeologists, the identification of 'what' professionals want to present, has become a highly complex issue. How can archaeologists identify and remove any messages that have 'gender bias, racial or ethnic preconception, and subjective evaluation based on degrees of technological processes' (Ucko 2000, ix)? An associated issue are the tensions between using chronological/cultural approaches to the presentation of sites and monuments and the demonstration of the veracity and limitations of the evidence driving the representation and interpretation of the past. The 'who' the presentation is for and 'how' it will be staged are also areas of lack of confidence amongst the archaeological community. With a wide spectrum of public interest and motivation, should visitors to sites be treated as developing thinkers with some competence and commitment, or should they be presumed to be of low competence and high commitment and therefore be directed and supported to acquire conceptual understanding of the monument (McManus 2000, xiv; Copeland 2006)? Reaching into the heart of the problem, Molyneaux (1997, 7–8) has commented that the technical language and concepts surrounding a visual source present few problems of understanding to the 'received and confident wisdom' of the 'insider', but text, both written in guidebooks and spoken in audio-tours presents an alien environment to the 'outsider'. Molyneaux (1997) suggests that rather than images being captured by text, professional archaeologists need to see these sources in this alien light to sense their visual power so as to confront the seen and represented world. This chapter aims to do just that and develops a framework for looking at low-visibility sites, perhaps the most difficult to access by the public, and the implications for the archaeological interpretation and individual's construction of the past.

Low Visibility Sites

Archaeological sites typically range from structures completely buried, of which there is no trace on the ground and that can only be detected by scientific processes such as geophysical techniques, to structures not visible but represented by earthworks ('humps and bumps'). However, it is the remains of only partly exhumed excavated structures that are the most visited by the public. Perhaps the most obvious example of this category of 'low visibility sites' in the United Kingdom are 'Roman remains', which usually occur in the countryside and are mainly of a military character. Though some villas and structures in urban areas are represented, pressure on land within towns – and the sequent occupation of these urban locations from the Romano-British period to the present – means that although excavation does take place, the recovered structural evidence is preserved by record and the site back-filled to be reused. In his influential publication, *Ruins: Their Preservation and Display*, M.W. Thompson (1981, 22) dismisses this type of visual source in just two sentences:

> The main difficulty with Roman remains is to *render them visible at all*, since they have to be laboriously revealed by excavation and then reset so as to be capable of withstanding the weather. If there are tessellated pavements of painted plaster, which cannot resist the weather, the only method of preservation for display is to roof them over, a costly operation that may lead to a disappointing result (1981, 22, author's emphasis).

However, since documentary evidence for the Roman period in Britain is scarce, we are confined to using the visual evidence of sites and artefacts from those sites, resulting in archaeology playing a larger role in reconstructing a history of the Roman period in Britain, whoever is constructing it (Mattingly 2006).

For a large number of people who visit sites like those being discussed here, the main stumbling block in understanding is that of the visual, and although the question 'Why did the Romans live underground?' is apocryphal, I did actually hear a primary school teacher on an in-service course at an archaeological site ask: 'Why did they knock it down and bury it before they went?' While undertaking fieldwork for this chapter, I experienced at Piercebridge Roman site in the North of England (see Figure 14.1) a small boy trying to look under one of the low remaining walls as if trying to lift up a carpet. He also stood quite still and bemused when inspecting one of the walls disappearing under a bank, which itself was topped by the modern site fence (his father thought this so amusing that he placed it on 'YouTube'). A major problem that is revealed by each of these examples is the lack of realization that the Romans themselves did not see these remains either, since they are largely comprised of the foundations of structures that would have been hidden beneath the ground, and part of the construction process. Where there had been rebuilding on different lines after temporary abandonment of a structure, both sets of walls are visually accessible to us when revealed by excavation, but the earlier set would have been buried beneath subsequent floors and therefore invisible in the following occupants of the building. What we see now are the foundations of structures, the results of stone robbing as well as the stripping of the layers of soil by archaeologists so there are no indications of the processes of dereliction, and usually no explanations of the

Figure 14.1 New Interpretations: Display Boards Create a Visuality Largely Missing in the Sites Themselves

Source: With permission of Linda M. Shovlin

organic and human agencies on these sites to demonstrate why what can be seen now is like it is. This is partly because, although 'hidden' places, they are also public and have been tidied up and remain as mathematical entities of straight lines and right-angled walls inscribed into the earth and, as such, in the present lack the contrasts of shapes, light or different textures that might indicate aspects of lived lives; function as it were, implied by form. While the excavation report and/or reconstructions of the remains would readily be admitted to the class of 'images' (Moser and Smiles 2005, 2), this type of site is usually excluded even if it is the result of human imagination after its disinterment. There are no clues for any other senses either: no sound of the running water that would have filled the sloping drainage system, or the Latin (or otherwise) of the legionary soldiers, nor 'the feelings of collectivity and solidarity, lost skills, ways of behaving and feeling, traces of arcane language and neglected historical and contemporary forms of social enterprise' (Edensor 2005, 167). Such ruins are sterile after excavation, 'almost geological structures' (Ball 2009, 385), the rooms covered with gravel to inhibit the growth of weeds. In ruins, instead of pre-arranged spectacles, the visual scene beheld is usually composed of no evident focal point but simply an array of apparently unrelated things. These are extraordinary and incomprehensible objects that are not commodities, obscure functions and sensations to assimilate (Edensor 2005, 167). However, to paraphrase Benjamin (1999, 576), just as the earth is the medium in which archaeological sites are buried, so the visual is the medium through which they are experienced.

Constructing Meaning at Exhumed Sites

How can we attempt to identify how these low-vision sites can best be unpacked so that visitors, both professional archaeologists and heritage tourists, can be helped to understand the significance of these remnants, the 'unloosing process' of Balm and Holcomb (2003, 160–2).

A number of processional models might be put forward, either from the educational world, as understanding an archaeological site is broadly a learning process, or by using visual methodologies and the interpretation of visual materials. Both approaches have similarities and differences but might be summarized as 'eyes-on' before 'minds-on': familiarity with the physical aspects of an image before making inferences. A conceptual vehicle for understanding such sites is provided by schemas produced by a number of scholars. Each involves a progressive sequence of familarization where visual understanding can be developed progressively either on a single site or on a linear trail, such as a Roman road, of low visibility monuments (Copeland 2006, 89). It is also interesting that each of these incremental models describe the process of disinterring a site through excavation.

Jerome Bruner (1971, 44–5) suggested that 'any idea or problem or body of knowledge can be presented in a form simple enough so that any particular learner can understand it in recognizable form' and this results in three modes of representation. An attempt has been made here to consider the implications of these in terms of archaeological sites. To start, Bruner introduces the *enactive mode*, which he describes as moving around the site, climbing walls, exploring the positive and negative features of the site such as ditches and embankments. He then moves on to what he labels the *iconic mode*, which, for Bruner, involves looking at pictures of reconstructions, re-enactments, museum displays of artefacts and so forth, and guided tours to make connections between the surviving evidence and possible interpretations. Finally, Bruner talks of the *symbolic mode*, which means understanding the context of the site in its past and present through maps, plans, and written language, and to understand that the whole is the subject of debate and shifting interpretations.

Liebeck (1984) proposed a chain of concept developments based on Bruner's work, insisting that *a store of visual images* (my italics) was needed on which to draw in order to form these abstract ideas. She identified experience, spoken language, pictures and the symbolic in terms of maps, plans, writing etcetera as the hierarchical order of progression. The significance of spoken language on historic sites has rarely, if ever, been explored as an adjunct of making meaning alongside the visual imageries. Likewise, Panofsky (1957, 26) suggested that the subject matter or meaning of visual forms is 'to be established by referring to the understandings of symbols and signs in a visual source that its contemporary audience would have had. Interpreting those understandings requires a grasp of the historically specific intertextuality on which meaning depends'. However, looking at low-visibility archaeological sites makes different demands on the perceiver than, say, looking at a work of art (a painting or sculpture or photograph), largely due to the amount of the site in three dimensions with an emphasis on the horizontal rather than vertical available to the observer and the general paucity of material remains. Because of this, the experience requires physical movement around the site to orientate oneself to appreciate the extent of the evidence from a 'collapsed activity', which are what archaeological sites comprise (Ingold 1993). Trying to categorize these actions is difficult.

Using Panofsky's progressive model, we can conjecture its applicability to low-vision sites. Close observation is demanded of the compositional interpretation in his 'primary stage/natural/pre-iconographic' stage, and this close observation would be facilitated by movement around the site, recognizing that there are stone features that are said to have been built by the Romans. These types of visitor were expected to know which was the inside or outside of the buildings and perhaps see the features as inscribed onto a site, rather than growing up through it. Panofsky's 'secondary/conventional/iconographic' phase involves using images that have a specific symbolic resonance might include (re-)constructions possibly (re-)enactments, movement around the site is usually decided by a pathway or way marked route indicating the detail to be seen, the special features of the site providing an economy of visuality. The modern routes do not necessarily correspond to those used in active life of the structure. Finally, Panofsky identifies an 'intrinsic/symbolic/iconological' mode of interpretation in order to explore 'the underlying principles which reveal the basic attitude of a nation (a county) period, class, a religious or philosophical persuasion qualified by one personality and condensed into one work', which seems to indicate understanding how sites are embodied in the landscape and the techniques by which they have been retrieved. Using archaeological sites makes this final part of the process very difficult as the individuals are anonymous and the production of sites was collective. However, the standardised plans of Roman fort/resses are Empire-wide and their lay-outs indicate the need to reflect power. Similarly, our knowledge of the functions and layout of Roman villas comes from Continental examples, especially the plentiful Roman literature on the Italian countryside. However, there were local variations suited to climate and terrain.

Panofsky's idea that '[i]nterpreting those understandings requires a grasp of the historically specific intertextuality on which meaning depends' has an echo in Giuliana Bruno's (1993) 'inter-textual montage' achieved through 'inferential walks' through Italian culture in the first decades of the last century, which she described in her study 'Streetwalking on a Ruined Map: Cultural Theory and the City Films of Elivira Notaria'. Since only fragments of Notari's films exist today, Bruno illuminates the filmmaker's contributions to early Italian cinematography by evoking the cultural terrain in which she operated by using an innovative approach: a critical remapping through the interweaving of examples of cinema with architecture, art history, medical discourse, photography, and literature with the panorama ranging from the city's exteriors to the body's interiors. The title and methodology of her work seemed even more apt when I decided to use the Dere Street Trail[1] as a case study for examining the imagery of Roman sites in the present, and the possibilities of 're-mapping' through walking along the route of the road, which itself is fragmentary in places, and examining the relationships between movement and the visual. Using Dere Street as the 'city exterior' and the individual fort sites as the 'body's interior', I hoped to see how it was possible – if it was – to access the cultural terrain of the Roman period, and through the visual remains of the Roman sites to see which methods were most effective.

The Dere Street Trail

There are problems in using Roman roads and their settlements as the focus of trails joining low-visibility sites. We are used to see Roman roads as being satisfyingly straight and direct, and thus easily identifiable on maps. It might be thought that there are advantages to using 'abstract' space so that traces of the road can be located in the totality of space. However, such maps create an entity which is 'independent of any point of observation (that could only be) directly apprehended by a consciousness capable of being everywhere at once and nowhere in particular' (Ingold 1993, 155). The road is represented as a visually linear feature with the only movement possibilities being forward and back, and return journeys in the present that are easily measured in Roman miles resulting in the calculation of the average marching distance per day needed between the roadside settlements. This is all valuable information and makes absolute sense to a life lived on a flat plane. Such two-dimensional representations record the *inscription* of the route of the road but are independent of being, and as such, this denies any process of movement through a landscape. Route maps, valuable as they are for the beginnings of landscape explorations, present a view that would have been completely incomprehensible to those who lived, and live, alongside and travelled, and travel, the road (Copeland 2009).

However, in contrast, if Roman roads are considered to be a three-dimensional entity *embodied* in the landscape then a *movement* of change and continuity, causation and outcome can be seen as being manifest in the visual and holds together the landscape. This embodiment in the landscape is the result of roads being seen not as built through abstract space, but through cultural spaces produced by people and their mobility within them (Witcher 1997). Such places have been described as 'place ballets' (Seaman 1995) where activities occurred in the past and which often resulted in visual evidence that has survived into the present – the low-visibility sites along the

1 Available at: http://www.durham.gov.uk/durhamcc/usp.nsf/pws/Archaeology+-+Archaeology-Projects-Dere+Street+Trail [accessed: 28 February 2009].

Figure 14.2 Walking the Road: The Dere Street Trail at Lanchester (Longovicium), County Durham

Source: With permission of Linda M. Shovlin

road. By moving along the route as a walker or motorist, a three dimensional map 'is organized around the passage of the traveller, and their perimeters are the perimeters of the site or experience of the traveller' (Copeland 2004), and is probably as close as we can get to 'telling stories' of a place as it was perceived by an individual or a culture moving through it' (Macfarlane 2007, 149).

Dere Street is the route of a Roman road whose contemporary identity is lost to us, though it is mentioned in the Antonine Itinerary (probably fourth century) as part of *Inter Britanniarum 1*, which joined Hadrian's Wall and York. The itinerary survived not as a map but as a manuscript, and seems to have been assembled from earlier ones. Although it names four of the forts, except Piercebridge, not all the names and mileages were transmitted correctly when copies were made in the Middle Ages – a 'ruined map'. The route of the road probably dates to around 80AD with the advance into Scotland by the Roman army, but the present remains of the forts are likely to be dated to the second and third centuries. For large stretches, the Roman road is followed by its modern successor, the B6275, before becoming a minor road beyond Bishop Auckland. There are a number of places where Dere Street diverges from the modern route and is represented by its *agger*, the raised foundations of the bed, which can be seen as a piece of 'land art' shadowing its descendant. In parts, the Roman road leaves the present line of the highway and crosses countryside, but this is not always followed by footpaths and in some sections its route is obscure or conjectured (Margary 1955). Possible lines are marked hedgerows, derelict sunken lanes hollowed out by rain-wash to the bedrock, terraces or hollows in fields and at one point it seems to have been destroyed by modern mining, all providing visual challenges to be solved, and an appreciation of the way nature actually affects a Roman structure. Georg Simmel suggested that the interest value in archaeological ruins could be traced to the way they reveal a contest between nature and culture, a proof that the cultural object (the ruin) can resist the ravages of nature (Wolff 1959).

The *Dere Street Trail* has been produced by Durham County Council Information Service as a project that forms part of an 'Archaeology in Durham' theme. Information about the trail is delivered in several forms: through a website devoted to its narrative and via information boards indicating what can be seen, or what is known about areas now under agriculture or modern occupation (see Figure 14.2). The trail's logo consists of an auxiliary trooper with an oval shield, one foot on a rectangular stone, which looks strangely like a ruined wall (and if so adds time-depth to geographical distance). The figure is standing at the junction of two roads, one of which enters a fort gate, the other, perhaps, being Dere Street, as it did not go through some of the forts but formed a by-pass. The roads on the logo are made up of large flagstones, although there has been little evidence for this in excavations carried out along the route. That this type of surfacing is associated with Roman construction is probably the result of popular representations of Pompeii. The soldier is on the main road, but looking into each fort as the present day tourist does – almost an invitation to view, but also a sense of movement between forts which is re-emphasized and consolidated by each fort having an information board with the logo on it, demonstrating an inter-connectedness and movement along it. The focus on the road as in-between places (Lawrence 1999) emphasizes the military associations, but civilian sites are also identified when near forts. If there were 'native' settlements on either side of the road then they are not included in the themed route and so an opportunity to present the ethnic mix along the route throughout the Roman 'Occupation'. Because the trail largely follows modern roads for much of its length, a car seems to be the intended mode for movement. If the marching style of the Roman army was to be used, although an enactive mode of following the route to experience the Roman experience of movement, it would not only take several days but would be an uncomfortable project along a busy road. My journey was undertaken partly by car and also on foot in the sections that are disputed or move across country.

The background to the trail is given on the website that shows the route of the road (although some of the illustrations are of forts not along Dere Street but on Hadrian's Wall). The 'carrot' is that the information on the map and site boards will also help you visit the Dere Street Roman forts and discover '"archaeology without digging", pits which contained sharpened wooden stakes, 28-seater toilets and the remains of Dere Street itself!!!', although the pits and the toilets sadly do not survive today. The main thrust of the interpretation is in the five forts along the route and they will be discussed from South to North along Dere Street. The trail begins at Catterick.

Catterick (Cataractoniu): The Town of the Waterfall

Catterick marks the division between lowland agricultural land and the beginnings of the higher land of the Pennines. It is here that the landscape of the road changes and this is reflected in the type of occupation from a thriving town with a villa(s) to a mainly military function reflected in the present by the British Army's live-firing ranges on the moors above and to the West. Today there is little to see except a depression alongside the trail's car park, perhaps indicating an amphitheatre, as the site is mainly under the racecourse. Excavations and aerial photographs have demonstrated that some 2 km to the South of the present town a roadside settlement developed alongside Dere Street within some ten years of the establishment of the Flavian fort in about 80AD. This settlement, which was entirely Romanized in character with rectilinear buildings from the start, appears to have reached its greatest extent in the mid-third century, but continued to be occupied into the fourth. Limited evidence has also been recorded for a probable villa within 400 metres of the roadside settlement, during the later third to fourth century. Three interpretation boards give the evidence from excavations, although nothing can be seen. Again, it is the military that is focussed on and not the important civilian settlement or the Neolithic/early Bronze Age henge on the same site. The board showing the possible Roman amphitheatre is complete with gladiators and in the excited crowd several women, again a copy of an Italian example.

Piercebridge (possibly originally known as Morbium)

The visual archaeological evidence is part of the Eastern defences, and these comprise of two linear discontinuous grassed over depressions in the ground that represent fort ditches and the now buried indications of sharpened wooden stakes, part of a courtyard building although the courtyard is under later structures, and an internal street. An aerial photograph, however, shows the crop marks of a small town, a *vicus*, with its roads and buildings. There is nothing to see more than four courses high. Since the place of the fort in the landscape and in relation to Dere Street and the possible small town is best observed from across the River Tees, an information board has been placed above and 400 metres away to give a panorama of the landscape (see Figure 14.1). It shows the typical plan of an auxiliary fort with the excavated area outlined, therefore developing a 'mini-trail' to the excavated and the now land-locked bridge, although above the site and 400 metres away on the opposite bank of the River Tees is a board with a plan of the existing remains. The position of the board is on Dere Street, which by-passed the fort and crossed the river via the bridge. The latrine building is illustrated but has been re-buried after excavation. A leaflet produced by Darlington Borough Council gives many details about the age, function and layout of the remains, but without a plan (although there is an advert for a local hotel!) presumably to be used in connection with the Dere Street information board. The tenuous nature of archaeological interpretation is emphasized by a statement that states that the information was accurate at the time of printing, but there is no guarantee that it will be in the future.

Ebchester (Vindomara)

The information board has a fence on it mimicking the real one in front of it and so giving authenticity to the view as well as orienting the visitor. Little remains to be seen of the fort, as the town of Ebchester was built directly on top of the fort, unlike the successors of many other Roman towns, which are generally situated at a little distance from the ancient sites. Ebchester, however, stands right upon the old site, and Roman ramparts, Roman altars, and Roman remains of all kinds are mingled in singular confusion with the gardens, cottages, roads, and St Ebba's Church. The trail board on the roadside includes an illustration of a train at an abandoned railway station which is now on a cycle route, the first indication that the area was heavily industrialized. This is emblematic of a change in the dominant history of an area, which also has a similar theme of Empire – the British and Roman. Edensor (2005) has commented on how industrial ruins, because of their fragmentation and alienation, are so different from the ordered urban landscapes around them. At Ebchester, the almost neurotic need for order by the Roman army contrasts with the ruins that now occupy the site.

Lanchester (Longovicium)

Since there is no public access to this fort, which is on private land, the site consists of a roadside information board and is located in a lay-by off the B6296 Lanchester to Satley road, West of Lanchester Village. The main theme of the board located at Lanchester is 'Archaeology without Digging', demonstrating the 'magic' of geophysics used to tell the narrative of the fort. The representation of the fort as it might have been is shown directly on top of the geophysics plot. Gateways, granaries and headquarters buildings are reconstructed with detail from elsewhere. Because the fort remains are so unprepossessing there is an 'Estate Agents' description of the site: 'One working fort set on rolling hillside in the bracing air of Britannia. Property comprising: and then the individual aspects of an auxiliary fort, barracks, watchtowers, headquarters ...' There is also an imaginative guided tour of the fort by Octavius Aurelius, Chief Medical Officer, although it is not clear whether there is evidence, in terms of a tombstone inscription, for example, that this individual actually existed. The Friends of Lanchester Roman Fort, a community based local history group, have produced the 'Roman Walk' – a circular route that does not pass through the fort. It does, however, use a mixture of *in situ* information and detail from elsewhere to 're-imagine' the structure of the fort. As a result of a Heritage Lottery Fund bid, the Friends have also been able to provide a series of information sheets. One of which asks the sort of questions used by professional archaeologist about age, purpose, garrison, layout, and providing evidence for each aspect: the geophysics plot, a dedication slab for the building of a bath (it was under the direction of Marcus Aurelius Quirinis, [who was a real person]), an altar and a building description of the construction by the XX Legion. Of all the forts along the Dere Street, Lanchester has the least to see but is the most fully interpreted.

Binchester (Vinovia)

Binchester is the most extensive of the sites along the trail with a section of a military bathhouse, a section of the commanding officer's and a short stretch of Dere Street are to be seen. It is the only site with a logo: an Imperial eagle and a border of the high status Samian ware pottery. A stone 'plan' of Binchester remains but there is a covering shed over the best preserved section of the baths The hypocaust under the military baths is on view as is a manikin of a soldier using a strigule an implement to clean oil from his legs and a re-construction of a wall with painted plaster.

There are a number of education activities connected directly with the site such as guided tours but also more 'generic' Roman activities for children found on a large number of sites and museums such as a mock excavation in a 'sand box' making a mosaic, replica costumes' which have little to do with what can be seen now, but as 'hands-on' activities' demonstrate the difficulties in providing a range of visual activities. However, '[f]or an extra charge and subject to availability it may be possible to book a *real* Roman Soldier to help bring your visit to Binchester alive' which is intended to make up for the paucity of the remains and the difficulties of representing and interpreting them. My own personal experience as an inspector of history in primary schools indicates that children are usually told that their field work visit will be to 'see the Romans' and without some sort of visual re-enactment are often disappointed and frustrated. Children's activity sheets are largely about 'collecting' parts of the site and 'Imagine that you are taken back in time and are a Roman soldier or local person at that place and write down words or phrase that describe what you would be able to see, hear, smell, feel', which again does not seem to rely on evidence from the site itself

Street Walking and the Visual

The Dere Street Trail gives a valuable insight into how people can be helped to construct the past. The available evidence and its interpretation at each fort on Dere Street produces not a 'circuit of memory' (Johnson 1995), but a 'linear remembrance' – all roads 'lead to Rome' and this was one of them. None of the forts has a lot of visual evidence that is accessible, with Lanchester having almost none. Binchester has the most 'standing' archaeology but does not necessarily have the most effective interpretation and representation of the period. The point is, however, that the whole route of the road can create 'memoryscapes' using iconographic forms as 'archaeological metaphors' (Edensor 1997). The active participation of following the route also involves experiencing its visual materiality. Each site has a different set of evidence and approaches to archaeology – excavation, geophysics, landscape study, and so it is possible to see how the whole trail might develop Bruner's or Panofsky's processional schema, whereas just visiting one of the sites without travelling the whole of Dere Street might offer only a fragmented conceptual experience.

In terms of my making a specific journey rather than simply using maps to make inferences about the road, it was necessary to undertake the enactive mode, which could lead to understanding the scale of vision from the road to individual sites and from the sites to Dere Street. This was particularly so in that my movement along the road gave me 'bearings' on sites from different points on the landscape as well as the distance and terrain along its course. The 'deep map' that was formed of the road had to be done by looking either side of it in order to explore 'the underlying principles which reveal the basic attitude of a nation period, class, a religious or philosophical persuasion qualified by one personality and condensed into one work' (Panofsky, quoted in Rose 2007, 151). Rather than Dere Street being a straight line, it is better compared to oscillations on a fluorescent screen reacting to the processes of the visual, memory, power and identity in the landscape.

The Dere Street Trail enables construction of pasts because 'as travellers, these social actors manifest their identity at moments of mobility through their active participation in the processes of movement' (Allen and Pryke 1999). However, the social actors might have different motivations and expect or need different ways of interacting with disinterred sites. While the overall impact of images on the site, the mood and atmosphere they engender might manipulate the viewers' reaction (Moser and Smiles 2005, 1) as important is the way that the viewer manipulates the site through their different motivations. Urry (1999, 39–40) has identified different types of visual consumption undertaken by a variety of audiences (Table 14.1).

Table 14.1 Different Types of Visual Consumption

Romantic	• Solitary, sustained immersion and sense of awe • Gaze involving sense of auratic landscape
Collective	• Communal activity • Series of shared encounters • Gazing at people who are also who are also familiar (on a site/along the road?)
Spectatorial	• Communal activity • Series of brief encounters • Glancing and collecting of many different signs of the environment
Environmental	• Collective organization, sustained and didactic, scanning to survey and inspect nature
Anthropological	• Solitary, sustained immersion, scanning and active interpretation

Source: Urry 1999

I have little doubt that I had an 'anthropological' approach to my consumption of the trail. No doubt the organizers of the trail will have an 'environmental' approach, but what about other audiences? However, there are some accounts of other who have undertaken the whole of the Dere Street Trail. A good example of the 'collective' type of visual consumption is provided by the log of the 'Multidaymen' cyclists, who had their Xth Anniversary Ride (or 'I Bungious rides South') from (Corstopitum) Corbridge to Eboracum (York) in June 2007. The web-contributor 'Bungious and His Legionnaires' rode down the Dere Street Trail, stopping at every fort and having their photographs taken by each of the trail interpretation boards. At Ebchester, the writer comments:

> ... it was here that we realized just how close we were following the Romans South ... To our own surprise we detoured from another cycle path to Binchester where it was closed 'but as had come so far we climbed the fence and surveyed the site. Well preserved and guarding the Wear as it was designed for ... Friday 29th June, Railways-Romans and more real ale ... Piercebridge camp and sunshine, we had a good stop and explored the camp plus down to the remains of the Roman bridge. Fascinating stuff. Bikes (Chariots) take a well earned rest ... Nice flat ride down to Catterick (another Roman site). At Catterick racecourse we discovered that in Roman times it was a site for bear baiting. A sporting arena for 15,000 years.[2]

2 Available at: http://www.kayamy.co.uk/2007intro%20anniversary_ride.htm [accessed: 28th February 2009].

Although they left the Dere Street Trail at this point, the penultimate photograph in the account is the statue of the Emperor Constantine in York, and the final frame that of a plaque on the pavement in Eboracum with a Roman legionary's helmet and the by-line: "Gateway to the Roman fortress': A good time was had by all!'

There are indications of a Romantic approach in the viewing of such sites as sacred (Tresidder 1999):

> Archaeology as a religious ceremony performed within the realm of the secular religion of the nation – that is Antiquity. Archaeologists act as the 'priests' of this religion, as mediated between the past and the present, while its monuments are its icons (Hamilakis 2007, 10).

It is easy to understand why archaeological sites might seem to be spiritual in some way: excavated structures disappear into the ground at some points, the interplay of revelation and concealment that is experienced as sites are walked around walked around makes them almost a form of land art, where they appear to emerge from the ground since only a part has been selectively uncovered due to the opportunistic or planned interventions by archaeologists. They appear to be immanent in the landscape – an energy waiting to be released. The archaeologist defines the area above ground that forms a liminal landscape between the visible and invisible, past and present (Tresidder 1999). While liminal places might be considered 'sacred' because the place is outside of time and space, the resurrection of it by the archaeologist adds to the mystery because they are trying to bring a place into time and space the religious/ritual ceremony of the guided tour or the missal of the guidebook. With the invention of geophysics and the ability to look below the ground without excavation, the role of the scientist as mediator between the present and the past is even more emphasized.

John Collingwood Bruce is especially remembered in archaeological circles as the man who started the *pilgrimages* to Hadrian's Wall, following the line of the Wall from East to West, 'forming a pilgrimage like that described by Chaucer, consisting of both ladies and gentlemen' (Burton 1989, 5), after which it was decided to hold pilgrimages every ten years: 'As this work in grandeur of conception, is worthy of the Mistress of Nations, so, in durability of structure, is it the becoming offspring of the Eternal City' (Breeze 2003, 9). As a religious minister, he may have used the Eternal City as a religious metaphor. The thirteenth decennial pilgrimage to study archaeological developments in the understanding of the Wall will take place in August 2009. Also in 2009, the World Community for Christian meditation will undertake a Hadrian's Wall Pilgrimage, and state that 'the countryside in that part of England is beautiful and we will be walking in silence for a large part of the day so it will be a time of real contemplative enjoyment'. This aesthetic, romantic, spiritual sensibility to ruins and historical sites has a long history and I certainly experienced feelings similar to these on the stretch of Dere Street where it crosses the countryside or its route was uncertain (Adler 1989; Woodward 2002).

So perhaps we must be looking for a number of visual roles in heritage visitors, especially in the 'spectatorial mode' which may well be the consumption pattern of the majority of visitors and for who the Dere Street Trail attempts to meet. It is the visual consumption pattern and the visual interpretation stage that needs to be managed if interpretation at a site or along a route is to be effective for a number of audiences. One of the most interesting aspects of the Dere Street Trail is the lack of visible evidence for the material structures at Lanchester, yet the plethora of interpretation. There would appear to be two main sources of interpretation, the Durham County Council Dere Street Project and the Friends of Lanchester Roman Fort, both with different visual interests and both presenting for different audiences' consumption patterns. No doubt the variety of consumption patterns of the Friends themselves reflects the material presented to the public. A further factor might well be that in order to qualify for the Heritage Lottery Fund grant the group received they would have had to demonstrate that the outcomes of the project were accessible to a wide range of visitors. This has resulted in a range of interpretation strategies that satisfy Panofsky's primary, secondary and intrinsic stages of visual methodology as well as a trail around the fort site also is an 'inferential walk' using the present and the Roman landscape to engender individual constructions of the past.

'Heritagescapes can reinforce the notion of a mythical place in which we can search for roots and authenticity' (Tresidder 1999, 138) and the latter are very important in the visual archaeological site. Mills (2003, 86) points out that it is 'in museums and heritage sites that the balance between high and mass culture has to be continually re-negotiated, for it is here that scholarship meets mass society'. He also notes that heritage venues are almost entirely focussed upon the provision of a sense of identity in terms of an essentially visual experience. In many cases, this visuality can be seen as un-academic as there is no chance for the visitor to contemplate that archaeological knowledge is always provisional and that it is retrieved in different ways that might produce different interpretations.

Visuality is an important part of archaeological sites as it is from these that the *in situ* evidence from the period comes. As seen above, it is the preservation of the sites that causes some problems, especially when the philosophy is not much different to that of John Ruskin and William Morris's 'conserve as found'. However, even so for many visitors it is the 'in this place', the *in situ*, that gives the greatest imprimatur of authenticity whether they understand the ruins or not. One of the attractions of watching a 'reality' television programme such as *Time Team* on British television is that the remains of structures can be seen as being made visual for the first time and are therefore authentic. The site at Binchester was the focus of one such episode, *Time Team – Street of the Dead: Binchester, County Durham* (Channel 4, first shown 13 January 2008), and this is quoted as further proof of the authenticity and importance of the site. This emphasis on the immediacy of the disinterring and the dialogue between archaeologist and the public, in this case in the form of the presenter Tony Robinson, demonstrates the provisionality of interpretations as more of the site is uncovered over the period of three days. This focus on the visuality of a site through an understanding of the techniques and the construction of the meaning by archaeologists may well be the way forward in moving representation and interpretation of sites away from just the historical and cultural mode.

The visual presentations along the Dere Street Trail are largely of the unexposed walls type, except for the simple visitors centre at Binchester and where that visual experience does not fully exist so that it can be quickly assimilated by the uninitiated visitor, subtle traces may have to be made more explicit with a need for balance between what is visible and the scholarly/academic. This is attempted along Dere Street by the provision of interpretive boards which do emphasize the visual in the presentation of information. Finally, to return to M.W. Thomson, individual Roman sites are not easy to interpret but they can be made visible by carefully identifying the types of visual consumption on sites, and in the case of inter-connected low visibility sites such as those along Dere Street, use the opportunity to develop visitors' ability to visually interpret the sites by an inter-connectedness of underlying concepts concerned with the presentation and interpretation of the past. This forcefully indicates the case for frequent evaluation on all the sites and continuous research into audience perceptions and visual consumption patterns and how they may be met.

From Frozen Monuments to Fluid Landscapes

Although Keith Emerick (1998) uses the visually evocative heading of this section as the subtitle to a chapter on providing physical access to medieval upstanding remains, it has many resonances for the problems of the visuality of the presentation of archaeological sites. The process of 'thawing' out individual low-visibility monuments may come from the mutuality of Urry's identification of visual audience, Panofsky's development of visual methodologies to meet the needs of that audience and the depth of Bruno's 'intertextuality' through inference trails. For example, the solitary and sustained motivations of Urry's 'Romantic' type of visual consumption would seem to be matched by Panofsky's primary phase. The inferential walk might be along the length of Dere Street, or on another foot-pathed route cutting through one of the sites. A professional archaeologist focussing on a site would be in Urry's anthropological area of visual consumption and would clearly be in Panofsky's iconological phase, constructing the landscape differently because of previously working through the previous stages of the schema by stint of a professional training. The mutual inference trail with high intertextuality would be not only the site, but the route of Dere Street and the development of the landscape since Roman times. The processes of discovery and preservation also would be part of his/her knowledge The fluidity of the landscape is achieved by the focus of movement along Dere Street which results in an understanding of the military strategies and values of the Roman army – Panofsky's Intrinsic/symbolic/iconological stage. Of course, the archaeologist can examine any site at the level of visual consumption that he/she wishes and with new sites move through different stages of Panofsky's continuum and also through increasingly complex inferential walks.

This is not the place to discuss particular techniques of presenting visual information as this had been done elsewhere (Copeland 2004, 138). However, it is hoped that this chapter has contributed to the debate on whether the beginning of interpretation strategies should be the archaeological text, or whether it is the visuality of a site framed by its grass surroundings, that is a powerful starting point for archaeologists' interpretation of low-visibility landscapes.

References

Adler, J. (1989), 'Origins of Sightseeing', *Annals of Tourism Research* 16, 7–29.

Allen, J. and Pryke, M. (1999), 'Money Cultures after Georg Simmel: Mobility, Movement, and Identity', *Environment and Planning D: Society and Space* 17, 51–68.

Ball, P. (2009), *The Sun and the Moon Corrupted.* (London: Portobello Books).

Balm, R. and Holcombe, B. (2003), 'Unlosing Lost Places: Image Making, Tourism and the Return to Terra Cognita', in D. Crouch and N. Lübbren (eds) *Visual Culture and Tourism*, 157–74. (Oxford: Berg).

Benjamin, W. (1999), 'Excavation and Memory', in M.W. Jennings, H. Eliand and G. Smith (eds) *Selected Writings*, Volume 2 1927–1934, 576. (Cambridge, MA: Belknap Press).

Breeze, D.J. (2003), 'John Collingwood Bruce and the Study of Hadrian's Wall', *Britannia* 34, 1–18.

Bruner, J. (1971), *Towards a Theory of Instruction.* (Oxford: Oxford University Press).

Bruno, G. (1993), *Streetwalking on a Ruined Map: Cultural Theory and the Films of Elvira Notari.* (Princeton: Princeton University Press).

Burton, R. (1989), 'Going to the Wall', *History Today* 39:9, 5–6.

Copeland, T. (2004), 'Presenting Archaeology to the Public – Constructing Insights on Site', in N. Merriman (ed.) *Public Archaeology*, 109–32. (London: Routledge).

Copeland, T. (2006), 'Constructing Pasts: Interpreting the Historic Environment', in A. Hems and M. Blockley (eds) *Heritage Interpretation*, 83–95. (London: Routledge).

Copeland, T. (2009), *Akeman Street: Moving Through Iron Age and Roman Landscapes.* (Stroud: The History Press).

Edensor, T. (1997), 'National Identity and the Politics of Memory: Remembering Bruce and Wallace in Symbolic Space', *Environment and Planning D: Society and Space* 29, 175–94.

Edensor, T. (2005), *Industrial Ruins: Space, Aesthetics and Materiality.* (Oxford: Berg).

Emerick, K. (1998), 'Sir Charles Peers and After: From Frozen Monuments to Fluid Landscapes', in J. Arnold, K. Davies and S. Ditchfield (eds) *History and Heritage: Consuming the Past in Contemporary Culture*, 183–96. (Shaftesbury: Donhead).

Hamilakis, Y. (2007), *The Nation and its Ruins: Antiquity, Archaeology, and National Imagination in Greece.* (Oxford: Oxford University Press).

Ingold, T. (1993), 'The Temporality of the Landscape', *World Archaeology* 25, 152–74.

Laurence, R. (1999), *The Roman Roads of Italy: Mobility and Cultural Change.* (Routledge: London).

Liebeck, P. (1984), *How Children Learn Mathematics.* (London: Pelican).

Macfarlane, R. (2007), *The Wild Places.* (London: Granta).

Margary, I.D. (1955), *Roman Roads in Britain: I. South of the Foss Way-Bristol Channel.* (London: Phoenix House).

Mattingly, D. (2006), *An Imperial Possession: Britain in the Roman Empire.* (London: Penguin).

McManus, P.M. (2000), 'Introduction', in P.M. McManus (ed.) *Archaeological Displays and the Public: Museology and Interpretation*, xiii–xvii. (London: Archetype Publications).

Mills, S.F. (2003), 'Open Air Museums and the Tourist Gaze', in D. Crouch and N. Lübbren (eds) *Visual Culture and Tourism*, 75–90. (Oxford: Berg).

Molyneaux, B.L. (1997), 'Introduction: the Cultural Life of Images', in B.L. Molyneaux (ed.) *The Cultural Life of Images: Visual Representation in Archaeology*, 1–8. (London: Routledge).

Moser, S. and Smiles, S. (2005), 'Introduction: The Image in Question', in S. Smiles and S. Moser (eds) *Envisioning the Past: Archaeology and the Image*, 1–12. (Oxford: Blackwells).

Panofsky, E. (1957), *Meaning in the Visual Arts.* (New York: Doubleday Anchor).

Rose, G. (2007), *Visual Methodologies: An Introduction to the Interpretation of Visual Materials.* (London: Sage Publications).

Rowland, T.H. (1974), *Dere Street Roman Road North: From York to Scotland.* (Newcastle upon Tyne: Frank Graham).

Seaman, D. (1995), *A Geography of the Lifeworld.* (London: Croom Helm).

Simmell, G. (1959), 'The Ruin', in K. Wolff (ed.) *Georg Simmell, 1858–1918: A Collection of Essays with Translations*, 259–66. (Columbus: Ohio State University).

Thompson, M.W. (1981), *Ruins: Their Preservation and Display.* (London: Colonnade).

Tresidder, R. (1999), 'Tourism and Sacred Landscapes', in D. Crouch (ed.) *Leisure/ Tourism and Geographical Knowledge*, 137–48. (London: Routledge).

Ucko, P. (2000), 'Foreword', in P.M. McManus (ed.) *Archaeological Displays and the Public: Museology and Interpretation*, ix–xii. (London: Archetype Publications).

Urry, J. (1999), 'Sensing Leisure Spaces', in D. Crouch (ed.) *Leisure/Tourism and Geographical Knowledge*, 34–45. (London: Routledge).
Witcher, R. (1997), 'Roman Roads: Phenomenological Perspectives on Roads in the Landscape', in C. Forcey, J. Hawthorne and R. Witcher (ed.) *TRAC 97 Proceedings of the Seventh Annual Theoretical Roman Archaeology Conference*, 60–70. (Oxford: Oxbow Books).
Woodward, C. (2002), *In Ruins*. (London: Vintage).

Chapter 15
Constructing Rhodes: Heritage Tourism and Visuality

Steve Watson

Tourism, according to Barbara Kirshenblatt-Gimblett, compresses life and eventually displaces it: 'organizing travel to reduce the amount of down time and dead space between high points', and, where it works through museums, historic buildings and other objects of the past, it 'escalates the process by which life becomes heritage' (1998, 7). Heritage is thus produced within dynamic processes of cultural production that select and present objects and embed them in the representational practices associated with tourism. As heritage attractions they in turn become the very media through which culturally important values are displayed and consumed, especially where these are associated with nation building and identity. The notion of consumption is particularly important here because of the tensions it creates between heritage in the form of museums, buildings and archaeology, and the visual modalities of interpretation and marketing that are central to the demands of the tourism industry (Allcock 1995; Dicks 2000, 180; see other authors in this volume).

Tourism thus provides a shop window within which are displayed those objects and aspects of a culture that support its important meanings and identities. As if to obscure this role, however, they are subject to a kind of aesthetic reductionism described by Vergo (1989, 48–50) and dependent on an objective authenticity of the type identified by Wang (1999, 353). Conventionally, attention is focussed on the object, its inherent value as such and its attendant typologies and taxonomies, rather than the processes that have made it significant. Frequently, heritage acquires the uses and values of a promotional brand in the consumption nexus that is the contemporary tourist industry (Waterton 2009; see Waterton this volume). In both cases, it is effectively deracinated and shorn of the socio-political meanings that provided the cultural context for its original production, a point made by Fraser MacDonald in his application of Debord's (1977) concept of the spectacle to the heritage of the Scottish Highlands: 'Selling history or heritage is contingent on the commodity being free from any association that could hinder capital accumulation … Selling heritage and place is therefore a highly selective business, which writes out or visibly excludes anything it cannot assimilate' (MacDonald 2002, 64).

I will argue here that the selectivity essential for the production of heritage 'commodities' is also a crucial mechanism in the role of heritage as an identity making process, an issue that has been under-researched until recent years

(McLean 2006, 3). Some authors, however, have begun to examine the deeper representational dynamics that reveal heritage tourism as an important identity-forming and essentially political practice. The centrality of these dimensions of heritage is well established. Allcock has put it succinctly, 'to speak of heritage is to speak of politics' (1995, 101) and Pretes (2003) has analysed tourism sights as a means for encoding discourses of nationalism and national inclusivity, a view that is central to Smith's (2006) comprehensive analysis of heritage as a cultural process. The significance of commercial and identity-making modes and discourses has also been recognized by Moser as a distinct way 'of participating in the process of meaning-making' in relation to popular and even professional understandings of the past (2001, 262–5). For Allcock, the politics of identity are central to heritage as a process rather than merely a collection of objects:

> By designating features of the past specifically as 'heritage', the items in question are endowed with an elevated status reminiscent of Emile Durkheim's concept of 'sacredness'. Heritage does not just refer to elements of our past: it designates things towards which we have an *obligation*. That obligation holds because the item in question is regarded as peculiarly important in defining an aspired identity. The identification of heritage therefore involves an attempt to create a sense of there being a moral bond which draws together a given community ... around the object in question ... and sets it off in significant respects from others (1995, 109).

The purpose of this chapter is twofold. First, I examine the significance of visual culture in these processes of selection and representation. I then move on to evaluate the contributions made by emerging theories of *visuality*, a term used here as both a metaphor for cultural significance and as a way of describing and analysing the processes by which objects are selected and represented in the production of heritage. This social visuality, I will argue, energises the representation of heritage through *agencies of display*, (Kirshenblatt-Gimblett 1998, 17–79), reflecting and representing important cultural referents such as the collective past, nationhood and identity. The result, which is encoded in the materialities of display, is a 'scopic regime', to apply Metz's terminology (1982, 61), in which the visual representation of places and objects as heritage is organized, managed and enhanced though essentially visual processes such as interpretation, visitor management and media content. In order to expose these processes, I examine them at work in the developing heritage tourism industry of the Greek island of Rhodes, which displays, at first sight, a diverse range of heritage objects from the Neolithic to the twentieth century. What is actually displayed in Rhodes, however, is more complex than the simple linearities of a heritage narrative. It is the product of an active cultural process that is organizing, reorganizing and representing space around selected objects of heritage and tourism, and in doing so, constructing a *Greek* island.

The data for this study was collected in two seasons of research on the island from 2007 to 2008. It involved detailed surveys of representational practice and touristic development at over 100 sites in all parts of the island at a time of considerable change. Touristic narratives were examined as they appear in websites, guidebooks, and signage, and in the tangibility of the island's past, in its tiny painted churches, castles, towns and villages.

Rhodean *Pasts*

At 80 km long by 38 km at its widest point, Rhodes is the largest of the Dodecanese islands and the easternmost of the Greek archipelago, situated in the Aegean Sea only 20 km from the coast of Turkey. Historically famous for the Colossus of Rhodes, one of the Seven Wonders of the ancient world, its contemporary cultural significance is assured by the status of the Old Town of Rhodes as a World Heritage Site. It seems likely, however, that cheap flights, package holidays, low-priced beach resorts and nightlife account for as much of its popularity with tourists as its undoubted heritage value. Despite the island's rich agriculture, tourism is now its largest source of revenue, with 80 per cent of local income generated directly or indirectly from the 1.5 million people who visit annually (General Secretariat of the National Statistical Service of Greece 2007). Tourism, however, is a seasonal industry and there is also something of a North/South divide, with tourism more significant in the North and agriculture still predominant in the South and Southern interior regions.

The history of the island can be divided into five more or less distinct periods, each of which has a particular relationship with Greek nationhood, culture and identity. Kirshenblatt-Gimblett, in a discussion of New Zealand tourism, has described such distinct phases as 'hermetic compartments' or cultural segments, formed from an imagined landscape projected by the development of tourism (1998, 141). Perhaps what they more clearly indicate is an *elective affinity*, to use Weber's term, between contemporary Greek identity and the island's past, the strength of which varies according to the particular period concerned and the extent to which it supports Greek national identity as a cultural imperative (Howe 1978).[1] The first of these periods is the island's pre-Hellenic prehistory, during which it was occupied in the Neolithic and early Bronze Ages by people who were central Asian in origin. The second is the late Bronze Age occupation of the island following the Dorian invasion of around 1100 BC. These Mycenaean Greeks established the island within the classical Greek sphere and linked it with

1 The concept of *elective affinity* was borrowed from the study of chemistry by Goethe to provide a metaphorical framework for a novel. Max Weber subsequently employed it to describe the mutual reinforcement of cultural meanings by association, most famously between the protestant religion and emerging capitalism in *The Protestant Work Ethic and the Spirit of Capitalism*, 1905. Howe's account of Weber's use of the concept has been influential.

the rest of the islands and the mainland epicentre of what eventually became an Athenian dominated culture (Torr 1885). The third period is the island's thousand year existence within the Byzantine Empire, during which its Greek Orthodox Christianity developed (Norwich 1998). The fourth is its European medieval period where, as an outpost of the Knights Hospitaller, it provided a base for their crusades in the Holy Land, and the fifth is the island's post medieval occupation as part of the Ottoman Turkish Empire (Kinross 1979). This lasted until the Italo-Turkish war of 1912 and the Balkan War of 1912–1913, which saw Greek annexation of the islands of the Aegean, except for Rhodes and the Dodecanese. These were occupied, supposedly on a temporary basis, by the Italians as a kind of ransom for the fulfilment by Turkey of conditions in the treaty that closed the former war (Schurman 1916; Kaldis and Lagoe 1979). After the Second World War it was briefly occupied by the British before being handed over to Greece.

Whilst these broad delineations obscure the complexity of the island's history, they do provide a basis for analysing and understanding the construction of its cultural heritage; in this case through the material remains that form the basis for tourism development and attraction formation. For Rhodes, the periods outlined above are differentially presented in ways that reflect their elective affinity with contemporary Greek identity as it is conventionally expressed on the island. Sites and objects are thus selected and represented not at random, but as objects that both reveal and express contemporary cultural and political concerns with its island and national identities. The content of heritage tourism as a representational process is determined, therefore, by a scopic regime that projects 'Greekness' onto the island's history and contemporary sense of itself.

Visual Culture and Tourism

Visual culture is at the core of tourism, which is, of course, essentially 'scopic'. Indeed, we are aware from the theorization of tourism (Urry 1990; Kirshenblatt-Gimblett 1998; MacCannell 1999; Crouch and Lübbren 2003; see Crouch this volume) that it is conditioned by visual culture, and even has its very own visual practice, known by that somewhat tautological term 'sightseeing'. This activity has been sanctified and naturalized in Western culture since the Renaissance, and received an important early cultural impetus from the Grand Tour, as Adler (1989) has described. The Tour was also, in itself, a major source of artefacts for eighteenth and nineteenth century museums and country houses, which, in turn, expressed the primacy of the visual in cultural consumption (Hooper-Greenhill 2000, 14–6). Whilst these activities seem far removed from the activities of the contemporary sightseer, there is a link, according to Adler, through the privileging of the visual and the role this has played in capturing a new visual domain and representing it perceptually as continuous and empirically knowable (1989, 24). By this means, the mysterious and mystical was demystified and controlled, colonized by applying familiar and conventional representations to the new and exotic lands revealed by travellers and explorers, and subsequently opened up by trade and empire building. David Bunn (2002, 148) has identified a similar motive factor in the emerging aesthetics of colonialism in Africa, where there was an 'expansion of a European aesthetic order into the world periphery during a period when capitalism was beginning to revolutionize its capacity to reproduce itself'. Similarly, Deborah Cherry has explored these processes in relation to the representation of Algeria in nineteenth century paintings. Here a pre-existing pictorial order imposed Western aesthetics and visual conventions, of which both recognized the otherness of Islamic architectural forms and at the same time forged them into a hybridizing encounter with the conventions of the colonizing power (2006, 54–6).

Whilst its effects in tourism have only begun to be researched, echoes of such pictorial ordering and recognition can be detected in the website imagery mentioned by Philip Duke (2007) in his examination of tourism on the nearby island of Crete. It seems that such visual conventions readily accrete around 'spaces' that, through the medium of tourism, become destinations. It is therefore hardly surprising to find it manifest and active as a central element of tourist attractions, a connection recognized by MacCannell (1999) in his schema for 'sight sacralisation', whereby an object, place or 'site' is transformed into a 'sight' for tourists. The penultimate stage in this process is 'mechanical reproduction', where sights are visually recorded and the images distributed through whatever visual media are available at the time, such as 'prints, photographs, models or effigies of the object which are themselves valued and displayed' (MacCannell 1999, 45). According to MacCannell, it is this stage in the process of sight sacralization that is most responsible for motivating the tourist to visit the 'real thing', and nowadays, this is amply supported by the visual cornucopia that is the internet. In the past, however, and to some extent still, the lithographic print, the picture postcard and the guidebook have all served, similarly, as markers of cultural significance, so that, as Peter Osborne (2000, 88) has put it, '[t]he Sight is tourism's essential object and location, and not only is the essential mode of consuming it visual, it is itself the result, the invention of predominantly visual representations'.

The mechanical reproduction and distribution of visual images provides a cultural marker of touristic interest and an authentication of that same experience. The presence of tourists is thus signified as valid and may be supported 'on site' by the paraphernalia of tourism, including signage, access, maintenance and conservation, information and the presence of ancillary services such as retailing facilities and hospitality, much of which use visual cues in their engagement with visitors. These considerations have focused interest on the production of markers such as guidebooks and picture postcards as deeper indicators of cultural production than their ubiquity and triviality might suggest, particularly where these can be read as visual texts (Mellinger 1994; Selwyn 1996; Dann 1988, 1996; Edwards 1996; Marwick 2001; Thurlow et al. 2005; Yüksel and Akgül 2007). In particular, Marwick (2001, 430) has illustrated the image and identity-sustaining qualities of postcards in her research in Malta, drawing a visual code together around the themes of otherness, the quotidian and the traditional, which ultimately 'help to confirm and validate elements and fragments of Maltese culture'.

The objects of heritage have been, and remain, a major source of tourist attraction and have played a significant part in supplying images for representational purposes, not only because heritage tourism is concerned centrally with the display of buildings and objects, but also with their interpretation for visitors, including tourists. Heritage is also centrally linked with specific constructions of the past that support both local and national identities and encode social and cultural meanings. As Philip Duke puts it, the latter 'continues to co-opt the past for its own purposes', and in so doing, has entered the contestation over meaning that characterizes contemporary uses of the past (2007, 14). Whilst the significance of 'visual interest' has been apparent since the early days of museum display (Hooper-Greenhill 2000) and the birth of the exhibition (Kirshenblatt-Gimblett 1998, 79–127), it is of heightened importance in contemporary culture because the visual is the primary sense through which attention is gained and maintained in an increasingly commercial, competitive and market-driven sector (Merriman 2004, 87). Costumed interpreters who demonstrate the use of objects and display technology are now an expected norm in the modalities of exhibition and interpretation, and recent developments in multimedia displays have heightened the visual in interpretive practices (Moser 2001, 267; Smith 2006, 199–200). Indeed, in some museums there is little besides such 'intangible heritage'.[2] Processes of interpretation have thus come to employ modalities of communication that depend, crucially, on 'visual interest' and even their own pictorial conventions that are 'laden with symbolic content' (Moser 2001, 266). In this they are aided with successive advancements in display technology and image reproduction, which combine processes of visualization with the need to stimulate audience responses (Brett 1996, 61).

For Levine et al. (2005), however, there is an inevitable tension between the cultural integrity of an object and the 'heritage narratives' that are constructed as a necessary part of the 'production of locality' that is needed as a basis for heritage tourism marketing. These narratives are dependent on signifiers of authentication or historical validation: 'a heritage destination must have both a sense of place and a compelling narrative to sell', something that archaeologists (for example) are increasingly expected to be complicit in providing (Levine et al. 2005, 401–2).

There are intimations in this discussion of a *visual* that goes beyond the *visible*. To follow such a line of enquiry might also help to resolve some of the debates about whether the ideas of MacCannell and Urry have placed too much emphasis on visual culture (see Meethan 2001, 81–9, for a discussion of these critiques). The idea of narrative construction in heritage implies an approach to display and to the visual that certainly goes beyond the mechanics of presentation, interpretation and effective marketing and enters that deeper realm of representational practice and agency, related to the issues of identity and culture mentioned earlier. As

2 The Museo del Baile Flamenco in Seville is typical of a museum that relies on enhanced visuality and performance rather than purely artefactual display.

Kirshenblatt-Gimblett (1998, 6) has suggested, visual interest is matched by 'vested interest'. Thus, 'display not only shows and speaks, it also *does*', and in doing so it orders and organizes its material referents in a way that not only sells attractions, but also reflects and affects the underlying meanings, identities, social structures and affinities that determine the society concerned. This display logic selects, presents and interprets objects, and places them in an ensemble of representational practice to create what Kirshenblatt-Gimblett calls 'hereness', (1998, 6, 78). Adding a temporal dimension to this analysis enables aspects of the past to be similarly privileged or excluded through the representation of material objects, artefacts, sites and places, particularly where these act in support of an elective affinity, a 'particular past' as Duke has termed it: sieved, selected and controlled by the agencies responsible for selling heritage to the public (2007, 22). Moser (2001, 276) has coined the term 'singularity' to express the way that such practices typically employ just one interpretation of past events at the expense of alternative perspectives or other stories. Duke identifies just such imperatives in his study of the touristic representation of Crete. The website of the National Tourism Organization, as well as other operators in the local tourism industry, 'prominently display ancient ruins or statues (clothed or naked) which occupy pole position with sun and beaches for alluring tourists' (2007, 23).

For Palmer (1999, 315), the selected images of the tourism industry in general, and heritage in particular, 'provide individuals with yet another means by which they can understand who they are and where they might have come from'. When these images and meanings are validated or authenticated by the official designations of the tourist industry and its governmental sponsors, the notion of the visual is expanded into a wider realm of meanings in a social, cultural and essentially political nexus: 'who is doing the defining, on what basis and for what purpose?' asks Palmer (2005, 8). Hooper-Greenhill (2002, 14–5) has used these extended concepts of the visual as a basis for examining museums as essentially visual discourses, and applies this as a critique of assumptions about the autonomy of vision and the objectivity of interpretation: 'looking is not a simple matter, and seeing is related both to what is known and to what counts as available to be observed. Seen in this context visual culture is working towards a social theory of visuality' (Hooper-Greenhill 2002, 14–5). At this level, I suggest, it assumes a metaphorical status, no longer merely a way of seeing, not even as the 'privileged sense', but a way of perceiving, of understanding and of knowing. Visuality as a metaphor describing social significance becomes both a source and a repository of cultural knowledge.

This might be seen more clearly in the representations and cultural uses of landscape. Its scale and relationship with nature lends it an air of permanence and the authority associated with endurance and stability. Yet, as Thomas has observed, '[i]n the Western world, landscape is predominantly a visual term, which denotes something separate from ourselves' (2001, 174). The English countryside, for example, becomes a cultural landscape based around imagery that evokes a deep past and the abiding sense of national character to which it gives rise so that '[a] landscape is a cultural image, a pictorial way of representing, structuring or symbolizing surroundings' (Daniels and Cosgrove 1988, 1). This imagery can be expressed with a few basic signifiers, including churches, country houses, thatched cottages, village greens and hedgerows, which appear whenever 'Englishness' needs to be invoked and made visible, or to produce symbols of national identity that can be used to attract tourists (Palmer 1999, 2005). Moreover, it embodies particular social structures and crystallises in its visuality issues of ownership, expressed in boundaries, hedgerows, and rights of way, so that as it is represented it is a symbolization of a material world that supports power relationships and class interests (Cosgrove 1983), something to which Rose has referred to as 'visual ideology' (1993, 89–101).

Visuality, as a metaphor for social and cultural significance, also provides a conceptual basis for understanding the role of culture in tourism. Heritage tourism, in particular, provides a context for understanding visuality in terms of the systematic and mediated processes of social construction that identify and represent important cultural referents such as identity and nationhood. In doing so, it denies the cultural autonomy of the object, the reducibility of heritage to aesthetics and the objective scrutiny of the observer, whilst at the same time revealing the vacuity of commodified representations. From a heritage perspective, it offers a way of seeing the past in cultural space, and the means by which heritage is literally framed, perceived, recognized and valued. Visuality, with its attendant representational practices and supported by the agencies of display generates a 'scopic regime', which may be either a 'harmoniously integrated complex' of visual practices or, alternatively, a 'contested terrain' of sub-cultural representations (Jay 1996, 3–4). Jay applied the concept to art history and particularly the effects of the perspectivalist revolution. With the notion of a culturally embedded visuality, it can also provide a unified and systematic way of understanding the visual in the service of, and in reciprocity with, the symbols of power, identity and nationalism that Palmer (1999) has already linked to tourism.

Visuality in the Heritage of Rhodes

It is ironic that the great monumental symbol of the Island of Rhodes, the Colossus that stood beside (not astride) the harbour of Mandraki, has not been seen for over 2,000 years and, indeed, only lasted 56 years until it collapsed during an earthquake (Karousos 1973, 40). Its actual appearance is unknown, because no contemporary images of it survive, but it does have a *retained visuality* preserved in its cultural significance and in the reproduction of old prints, such as the engraving by Philip de Bay in Fisher von Erlach's *Plan of Civil and Historical Architecture*, published in 1721. Such images typically represent the Colossus as an impossibly large statue exaggerated in its proportions by its status as one of the seven wonders of the ancient world. Its visuality is now expressed in the sale of souvenir statuettes and statues, which adorn tourist centres such as Falaraki as well as in restaurants and other public venues (Figure 15.1).

Figure 15.1 The Colossus of Rhodes Reduced to a Garden Ornament in a Restaurant in Rhodes Old Town

Source: With permission of Steve Watson

The destruction of the Colossus also stands as a metaphor for the burial of the island under an avalanche of contemporary tourism representations, and a visual culture that is dominated by the leisure and pleasures of its beach resorts. In presenting the island's heritage there is, therefore, an immediate representational dissonance that is carried by the need to diversify the tourism product and thus engage and include heritage resources in the tourism portfolio. The 'other' Rhodes is an agriculturally productive and varied landscape of fertile valleys, farms, wooded hills and rugged mountains. It also possesses the kind of heritage that is conventionally described as 'rich', with a World Heritage site in the Old Town of its capital, the Acropolis of Lindos and a variety of interpreted sites relating to its ancient Hellenic past and the medieval occupation of the island by the Knights Hospitallar. At another level, the countryside in the South continues to support communities for whom the tourism of the North and West of the island means very little, but which has begun to enter their own places and spaces with a proliferation of visual cues, signs and signifiers. These intrusions effectively redefine places and represent them in ways that local residents have not previously known so that the lost places of the past might, by these means, be duly 'unlost' (Balm and Holcomb 2003). Modulating this process are the elective affinities and relative cultural positioning of the historical periods briefly outlined above. The visuality of heritage on Rhodes is thus conditioned by the relationship of objects, places and historic sites with Hellenism and the imagined sense of Greekness that underpins its national identity.

The Classical and Hellenic

A few kilometres from an unmarked and largely neglected Neolithic site at Stelies are the fragmentary remains of a Lycean tomb. Although strictly speaking this is from the Persian invasion of the islands in the fifth century BC, its presence expresses a clear relationship with Greek culture. The remains, though slight, are clearly classical in appearance and associated with a time when the native Rhodians were resisting Persian aggression. They can thus be 'read' as an indicator of the formative period of Hellenism when a Greek army from Athens defeated and expelled the Persians from the island. Not only is this marked on the tourist map of the island, its presence is indicated on the ground by a new brown sign. As an attraction, however, it is somewhat wanting. Its vague outline apparent in a rock face is only visible from the car park of a nearby bar. Closer access to it is barred by a high fence, although there is actually little else to see. For Vergo (1989, 54) this might constitute a 'reticent object' dragged into the public gaze, but the mere visibility of the monument becomes an irrelevance as its visuality is assured by its cultural significance and enhanced by its signification on the map and by the brown sign. Thus officially authenticated, the tourist visits, sees and experiences the constructed visuality of the object as an expression of its cultural significance, through what MacCannell has described as 'marker involvement':

'The truth is that marker involvement can prevent a tourist's realizing that the sight he sees may not be worth his seeing it' (1999, 113). Kirshenblatt-Gimblett makes a similar observation in her analysis of the 'overcoded signifier': 'The signified may indeed turn out to be trivial, or it may take a lifetime to fully discover and understand what [is hidden]' (1998, 255). The marker thus ascribes aura to the object through its signification of cultural value and a metaphorical visuality that is then realized by signage and interpretation on site. It also draws our attention away from what is not seen or understood, and encapsulates our engagement with the site in the official narrative content and significance ascribed by the marker and any associated information. The information becomes a proxy for the object, 'creating interest in places, especially those lacking in noteworthy visual attributes' (Kirshenblatt-Gimblett 1998, 167). Our presence there is thus validated by the cultural significance of the object, a metaphorical visuality realized in the map and enhanced in the signage.

Other classical and Hellenistic sites are more substantial and have benefitted from conservation measures over a long period of time. The early Mycenaean period is represented by the settlements and monuments at Kamiros, Ialyssos, Rhodes Town and, most famously, at Lindos, although the latter may have originally been the least important (Karousos 1973, 108). The origins of these sites form a link with the founding myths of Mycenaean culture when, in the tenth century BC, the Dorian conquest established their religious significance for the next thousand years (Boardman et al. 1986). They are also associated with recorded history, religious cults and even individuals, who distinguished themselves in war, politics or the arts. At Kamiros, the Greek classical remains are much more extensive and visible, and have been excavated and conserved so that the site is one of the most significant attractions on the island.

Lindos (Figure 15.2) is very much the jewel in the crown as far as the Greek heritage of Rhodes is concerned, with a foundation myth that associates it with Tlepolemos, the son of Heracles, and a participant in the wars against Troy (Archontopoulos n.d., 14–6). The great rock of Lindos dominates the Eastern coast of Rhodes forming a promontory of some 116 metres in height, towering over the harbour and the town that has grown around its base. Whilst there is evidence that the site was occupied in the Neolithic period, most of the visible remains are of Dorian and Hellenic origin, overlain with buildings and fortifications associated with the Byzantine period and the Knights Hospitallar. The site was notable, however, for the cult of the Lindian Athene, a hellenisation of a pre-Greek goddess of unknown provenance (Karousos 1973, 109). Thus was established the acropolis, over which a temple was built and variously altered and rebuilt in accordance with the changing tastes of classical architecture. The building reached its zenith in the later Hellenistic period around 200 BC, as a sophisticated complex of temples and ancillary buildings covering the whole of the summit of the rock.

In modern times, the acropolis of Lindos has been extensively excavated and conserved, notably by Blinkenburg and Kinch of the Carlsberg Institute in Copenhagen between 1902 and 1914, although further excavations were carried

Figure 15.2 First Sight of Lindos, a Well-Established Viewpoint That Has Become an Attraction in Itself

Source: With permission of Steve Watson

out in 1952 (Sorensen and Pentz 1992). Moreover, it is a major heritage tourist attraction, second only to that of the Old City of Rhodes itself. The temple of the Lindian Athena, which crowns the site, was partly reconstructed during the Italian occupation (1912–1945) constituting what might be considered an extreme form of visual enhancement. A similar reconstruction was carried out at the stadium outside Rhodes Town under the auspices of the Italian Archaeological School. The reconstruction at Lindos is now being 'restored' as part of a long-term programme to recover what seems to have been something of a botched job, so that by now the investment in visuality has more or less replaced any concept of the visible remains that existed before the Italian works commenced.

It is difficult to avoid the conclusion that what was perceived and represented was a Greco-Roman common heritage and that following several centuries of Ottoman Turkish rule this was a statement in stone and concrete that the island had been returned to its rightful heirs, who drew both identity and political inspiration from the classical world. Initially with Greek help, the Italians established a colony, the *Isole Italiane del Egeo*. They demolished houses and cemeteries that had been built during the Ottoman era, whilst preserving the remains of the classical, Byzantine and Knights' periods, and established an institute to study the history and culture of the island. The enhanced visuality of the Stadium at Rhodes Town and the Temple at Lindos are thus more than a statement of their dubious qualities as restorations, and a clear reference to their historical-cultural significance as monuments to Greco-Roman civilization. Visuality as a metaphor for cultural significance is realized in concrete and stone, made visible for both tourists and indigenes.

Byzantium and the Greek Orthodox Church

The Byzantine Period begins with the division of the Roman Empire into its Western and Eastern parts at the end of the fourth century. An Eastern capital was established at Byzantium by the Emperor Constantine, after whom it was renamed Constantinople. But even the term *Byzantine* is a later construction and would not have been known to those to whom it is applied. It was known, however, as the 'Empire of the Greeks', and this, as well as its Eastern Orthodox Christian associations and its resistance to incursions by various Persians, Arabs and Saracens, may be significant in the way that its material remains are now treated on the island: for while the built and archaeological remains of this period are fragmentary they are well mapped and signed on the ground. This may, however, be an easy *post hoc* rationalization, given that the Byzantine, as Emerick (2003, 170) has made clear, was not always so well favoured in Greek island communities such as Cyprus, for whom the classical past had greater meaning in expressing their affinities with the Greek mainland. This was to change, however, during the twentieth century, as European academics and, significantly, tourists, began to take an interest in Byzantine art and buildings, and the Orthodox Church began to recognize its value as a historical and identity-forming cultural link with the mainland (Emerick 2003, 212).

Churches with Byzantine remains, often indicated by the presence of mosaics and Basilica such as at Agios Nikolaos Fountoukli near Eleousa and the Monastery of Metamorfossi in Asklipio, are particularly valued, and are often staffed by caretakers who are assiduous in their policing of bans on photography within the buildings. South of Apolakkia on the Western side of the island is the partly excavated site of the monastery of St Anastasia Zonara, slowly sinking back under the ground, though keenly marked by brown signs and the tourist map. Fragments of Byzantine marble columns and other architectural features often find their way into much later buildings, set in walls or used to support small altar tables, or are simply scattered around outside. They seem to act as a kind of talisman linking the present building to a distant ancestry that is emphatically Greek. Thus, the ruins and fragmentary remains of St Anastasia and the more complete churches of Agios Nikolaos Fountoukli, the Monastery of Metamorfossi in Asklipio, the church of the Dormition of the Virgin in Mesanagros and the Monastery of Skiadi have provided the prototypes for a continuing tradition of church architecture and decoration.

The 'painted churches' of the island have thus assumed something of an iconic quality, one that has not been overlooked by the embryonic heritage tourism industry. These churches and many others are well marked as attractions, and have effectively enabled the touristic representation of almost every church on the island that exhibits the visual conventions of this church-building tradition. This includes many recently built churches such as the Church of the Virgin Zoodochos Pigi near Apolakkia which, with an extravagant disregard for age as an essential component of heritage, are also considered to merit brown signage. A genuinely ancient church a kilometre or so away, however, is not distinguished in this way and survives only as a derelict shell: heritage sacrificed to the more pressing need for the reservoir that laps around it. Even iconic and identity-making culture has limits to its value, especially if there is a threat of drought. This is more than a little supportive of Yaniv Poria's point (this volume) that heritage spaces can exist without including ancient objects and that the latter are not always represented as heritage spaces. Otherwise, the tradition continues as a vestige of Byzantium and the persistence of Orthodox belief as culturally vital, in the freshly painted frescoes and icons of many new churches, represented as heritage as soon as they are built. That this is a heritage in active communion with the present was apparent in my discussion with the caretaker of the Church of Agios Giorgios at Sienna who commended her son as an excellent painter of frescoes and icons, while in a back street of the Old Town in Rhodes an artist's studio is dedicated to the production of frescoes and icons, for new churches and tourists alike.

The Knights

The medieval period on Rhodes belongs to the Knights Hospitallar, the Knights of the order of St John of Jerusalem, although this was the Knights in retreat from the failure of the Crusades and, in particular, the fall of Accra in Palestine in 1291. After first seeking refuge in Cyprus they were sold the islands of Rhodes, Kos and Leros by the Genoans, and by 1309 they were established in the island. The occupation by the Knights lasted until 1522, when they were ousted by the Ottoman Turks, but in the intervening two centuries they had left a legacy of building that contributes much to the island's emerging heritage tourism industry. Prime amongst these buildings is the old City of Rhodes itself, which was granted world heritage status in 1988. The town is dominated by a circuit of strong walls and ramparts with towers, gateways and outworks extending around its harbours. The town also contains the famous Street of the Knights, which housed the inns and hostels occupied by the various nationalities represented in the garrison. This street is now the touristic core of the town and one of its main attractions. There are also other public buildings and several churches, some of which are now ruinous, and, as a centre piece, the heavily restored Palace of the Grand Master (Kollias 1998). The overall effect is of a medieval town, although this tends to obscure each of the other of the island's historic phases that are also represented to varying degrees.

Apart from the spectacular buildings in Rhodes Town, the Knights also restored a number of Byzantine fortifications and built new castles, some of which are now entering touristic space as monuments, mapped and marked as part of the island's rich medieval heritage. Those that have received most attention, quite naturally, are castles that are relatively well preserved and easy to access: Kritinia and Monolithos in the West, and Asklippio and Archangelos in the East. Evidence of their new life as tourist attractions lies in a scattering of brown signage, conservation measures, postcards and tourism brochures, as well as visitor management such as the provision of car parks and ancillary services. At Kritinia, a ticket seller's booth has recently been installed and walls have been somewhat haphazardly repaired. At Monolithos, the precipitous cliffs upon which the castle is perched pose a challenge for health and safety, but this has not prevented easy access by a maintained path from the public road and a café bar at the base of the hill that draws its trade from castle visitors.

The impression offered by this example is that such developments are spreading through the castles of the island, meeting resistance only where the remains are too slight to excite interest or where access is, at the moment, too difficult to be addressed with a minimum capital investment. Thus at Feraklos, for example, the scale of the castle's remains are insufficient to counter its relative inaccessibility and the investment that would be required to consolidate its ruins and present them effectively for visitors. The site also poses considerable dangers in its fragile masonry, deep empty water cisterns and dense undergrowth. Despite its visibility, therefore, it seems unlikely to enter touristic space in the ways that other castles have. Similarly, the remains at Sianna, though known to archaeology, are sparse and lacking in access and presentation, a site that is literally not prepared for tourists. At Kremasti, on the North West coast, the boundaries of touristic space seem to dissect the castle itself. On the summit of a hill overlooking the town this characteristically Venetian castle presents a well preserved frontage, vaults with new paths and the church of St Nicholas attached to it which is well kept and normally open to the public. On the other side, however, the remains of the castle appear to serve as storage for the recycling of old rags and may even provided accommodation for the proprietor of this enterprise. At the back of the castle its outbuildings seem to spill down the side of the hill and dissolve into the houses and cellars of the surrounding town.

The cultural significance of the medieval heritage of Rhodes might have suffered the same ambivalence as occurred in Cyprus, where, until it was celebrated under British colonial rule by individuals and groups more used to valuing medieval art and architecture back home in England, it was considered part of an alien and invasive culture (Emerick 2003, 204). In Rhodes, however, the cultural significance of the Knights, their visuality and therefore their potential for projection in touristic space is guaranteed by contemporary touristic interest in the medieval, the world heritage status of the Old Town of Rhodes, and, not least, their symbolic resistance to the Turks. Whilst the Knights were not Greek, they were at least Christian and a source, therefore, of affinity with contemporary cultural imperatives. Yet the castles of the Knights pose something of a dilemma for the development of heritage tourism on Rhodes. Some of the sites are inaccessible, dangerous and difficult to manage as attractions, and some of those that are accessible remain hazardous for visitors arriving in numbers because of their instability and situation. Some of them will doubtless be propelled into a touristic space and their significance will be realized in visual enhancements, conservation measures, embedded interpretation and visitor management. Others, by contrast, will remain resolutely invisible, their visuality unrealized because of their inherent unsuitability as tourist attractions, their status as heritage compromised only by the trouble, expense or sheer impossibility of presenting them as such.

The Post–Medieval Period

The island's post-medieval history, from the arrival of the Turks in 1522 through to their departure in 1912 and the subsequent Italian occupation, is the least well represented in its heritage. The period began with the loss of the island to the Turks and the withdrawal of the Knights Hospitaller to Malta. The Turks, however, seemed to have concerned themselves little with the activities of the Greeks and their religion in the island as a whole, preferring to confine themselves and their political control to the City itself, which was barred to Greeks. In this state it existed for nearly 400 years until 1912. Between 1918 and 1923 various moves were made to have the island returned to Greece, but when Mussolini came to power there began a period of ferocious suppression in which land was granted to Italian colonists and farmers. Intermarriage with Greeks was encouraged as a means of assimilation into a Latinized regional identity (Doumanis 2005). Successive Italian Governors introduced still harsher suppressive measures until 1936 when Italian was decreed as the language of education and public life and the Greek Orthodox Religion was discouraged. All the while Greeks were using their limited freedom to emigrate to Western Europe and the United States, from which they formed a political lobby to press for the Hellenic cause in the Dodecanese. The Axis powers occupied and strongly fortified the islands during the Second World War and these were not given up until the general surrender in May 1945. After this, an interim government was set up by the British, with the islands returned to Greece formally in 1948. An otherwise sober account of the island's history records the Italian period with bitterness:

> The sufferings of all the islands of the Dodecanese including Rhodes, under the Italian rule are well-known. The methods already in use for the suppression and uprooting of the Greeks became more systematic, more complete, more cruel, under the fascist regime … At the end of World War II, the 'nymph of the sun' regained its freedom and was returned to Greece (Karousos 1973, 27).

The desire of the Italians to Latinize Rhodes resulted in the restorations and reconstructions of Hellenistic sites discussed above, primarily in Rhodes Town and at Lindos. In the former there was a wholesale demolition of Ottoman houses and other buildings whilst those of the Knights period and the remains of Hellenistic structures were consolidated and restored. The main result of this is the relative 'silence' of the Ottoman period in the island's heritage. Whilst the Suleyman Mosque is a major building in the centre of the Old Town, it is not an attraction as such, opening being normally restricted to worship.[3] Nearby, the library of Hafız Ahmet Ağa is open to the public, but with little to commend it to any but the most inquisitive of tourists. South of the mosque, the Turkish quarter retains its old street pattern around the Platia Arionos, were some of the mosques have long been converted into Christian churches. The Yeni Hamam bath house and the remains of public fountains have some small interpretation plaques that appear to be related to World Heritage status but there is little else in the way of signage or obvious care for the fabric of the buildings. To the North of the Old Town is the Muraid Rais Mosque, named after an Ottoman admiral, killed in battle, and whose tomb it now contains. The mosque is surrounded by a Turkish cemetery much favoured by the town's cat population, and in its grounds is the villa occupied by Lawrence Durrell when setting up the island's newspapers during the post-war British occupation. His reflections formed the basis for a slight but well known travel book that is replete with classicist sentiment and not a little anti-Islamic (Durrell 1954).

These neglected buildings ironically offer much in atmosphere and the patina of age possessed by a heritage that is not yet 'managed'. Things may be about to change, however. The Deputy Mayor of Rhodes, for example, has recently announced that the Ottoman heritage is 'an inextricable part of Rhodes History' and has thereby instigated a programme of restoration. An historian on the project made the revealing claim that 'this work is done with love and detail, regardless of whether we deal with Ottoman monuments' (Manis 2008). Whether or not this move is successful in reintroducing the buildings to public life, as Manis puts it, is yet to be seen and whilst the effort is clearly progressive, its very necessity is an indication of the situation hitherto. Clearly, however, this heritage must have gained some cultural significance for its visuality to be realized on the ground.

Conclusion

The heritage tourism of Rhodes is a broken mirror to its history, reminding us, as Lowenthal (1998) pointed out, that these two ways of representing the past are necessarily different, and that the role of tourism in differentiating them is both powerful and pervasive. Tourism is employed centrally in representing and displaying the Greekness of Rhodes, in contrast with other aspects of its history which are at variance with this narrative. The island's distinct historical themes therefore stand in distinct relation with this 'particular past', and the nature of this relationship determines their cultural significance or metaphorical *visuality*. This visuality is realized through what is made actually visible in the various media

3 Although it was closed completely at the time it was visited for this research.

that represent Rhodes and its past. Other objects and pasts are, conversely, denied this actuality and whilst as objects they may be seen, their lack of visuality reflects their lack of cultural significance and the privilege of projection in the island's heritage tourism.

By the same equation, the Colossus of Rhodes, that symbol of the island's mythical past, is invisible, thanks to an earthquake in antiquity, but yet it retains a visuality born of its cultural significance, which makes it available for purchase as a statuette in souvenir shops, on postcards or as an element of interior décor in a restaurant. What we see as tourists in Rhodes is ultimately the product of a scopic regime, a logic of display, which at once constrains and expresses the island's sense of identity and self, and it's constructed past, through the essentially visual media of heritage tourism.

Visuality as a metaphor for cultural significance is an essentially socio-political process operating through the nexus of tourism and employing touristic representational media and sign systems. As such, its objects are not necessarily visible, although their cultural significance may make them so through tourism. Through this spatial visuality, the past becomes manifest, communicable and reproducible as artefacts, sites and places to visit and see, and things to consume. When these processes are combined within the representational nexus that supports heritage tourism, they also effectively construct and promulgate an authorized visual narrative, the visual equivalent of an authorized heritage discourse (Smith 2006).

The link with tourism also effectively reduces the display of cultural objects to an exercise in branding, and thus a part of the promotional activity that is central to the marketing of the island as a tourism destination (see Waterton this volume). Brands, however, can only be challenged for their value as such, for their efficacy in representing the island to the outside world, and to itself. A whole raft of cultural objects and their selective representation is thus removed from the political nexus wherein it might be challenged and modified. The significance of this cannot be over-estimated, especially given the pervasive strength of the internet as a medium for promoting destinations and, in effect, defining them for touristic consumption. These representations are, in turn, contingent upon the same objects having been previously reduced to aesthetics, objects without a past other than that which is constructed in the moment of recognition in their cultural domain.

The scopic regime of Greekness, and Greekness in the face of other competing narratives, is thus naturalized by the representational processes and marketing modalities that support it, together with an organizing aesthetic that reflects a certain hegemonic power and authority. The result is a particular past and 'hereness', a Greekness revealed and emphasized, a singularity of identity that is privileged in relation to the abiding 'others' of the island's past. Herein lies the value of visuality as a means of understanding cultural processes that are otherwise paradoxically obscured by what is made visible in the construction of place, identity and product that is central to contemporary tourism. It is perhaps unsurprising to find in a plan to rebuild the Colossus of Rhodes, that Falaraki rather than the old town of Rhodes itself has been chosen for the site (BBC 2005); the most significant element of the island's heritage made visible again in its newest tourist resort.

Bibliography

Adler, J. (1989), 'Origins of Sightseeing', *Annals of Tourism Research* 16, 7–29.

Allcock, J.B. (1995), 'International Tourism and the Appropriation of History in the Balkans', in M.F. Lanfant, J.B. Allcock and E.M. Bruner (eds) *International Tourism: Identity and Change*, 100–12. (London: Sage).

Archontopoulos, T. (undated), *Lindos Archaeological Guide*, (Athens Adam Editions).

Balm, R. and Holcomb, B. (2003), 'Unlosing Lost Places: Image-making, Tourism and the Return to Terra Cognita', in D. Crouch and N. Lübbren (eds) *Visual Culture and Tourism*, 157–74. (Oxford: Berg).

BBC (2005), 'Rhodes Plans to rebuild Colossus', [Online: BBC News Channel]. Available at: http://news.bbc.co.uk/1/hi/world/europe/4264169.stm [accessed: 1 September, 2008].

Boardman, J., Griffin, J. and Oswyn, M. (1986), *Oxford History of the Classical World.* (Oxford: Oxford University Press).

Brett, D. (1996), *The Construction of Heritage.* (Cork: Cork University Press).

Bunn, D. (2002), '"Our Wattled Cot", Mercantile and Domestic Space in Thomas Pringle's African Landscapes', in W.J.T. Mitchell (ed.), *Landscape and Power*, 127–74. (Chicago: University of Chicago Press).

Cherry, D. (2006), 'Algerian In and Out of the Frame: Visuality and Cultural Tourism in the Nineteenth Century', in D. Crouch and N. Lübbren (eds) *Visual Culture and Tourism*, 41–58. (Oxford: Berg).

Crouch, D. and Lübbren, N. (eds) (2003), *Visual Culture and Tourism.* (Oxford: Berg).

Cosgrove, D. (1983), 'Towards a Radical Cultural Geography: Problems of Theory', *Antipode* 15, 1–11.

Daniels, S. and Cosgrove, D. (1988), *The Iconography of Landscape.* (Cambridge: Cambridge University Press).

Dann, G. (1996), 'The People of Tourist Brochures', in T. Selwyn (ed.) *The Tourist Image: Myths and Myth Making in Tourism*, 61–82. (London: Wiley).

Debord, G. (1977), *Society of the Spectacle.* (Detroit: Black and Red).

Dicks, B. (2000), *Heritage, Place and Community.* (Cardiff: University of Wales Press).

Doumanis, N. (2005), 'Italians as "Good" Colonizers: Speaking Subalterns and the Politics of Memory in the Dodecanese,' in R. Ben-Ghiat and M. Fuller, (eds) *Italian Colonialism*, 221–32. (New York: Palgrave Macmillian).

Duke, P. (2007), *The Tourists Gaze, The Cretans Glance.* (Walnut Creek, CA: Left Coast Press).

Durrell, L. (1954), *Reflections on a Marine Venus: A Companion to the Landscape of Rhodes.* (London: Faber and Faber).

Edwards, E. (1996), 'Postcards: Greetings from Another World', in T. Selwyn, (ed.) *The Tourist Image: Myths and Myth Making in Tourism*, 197–221. (London: Wiley).

Emerick, K. (2003), *From Frozen Monuments to Fluid Landscapes: The Conservation and Preservation of Ancient Monuments from 1882 to the Present*, unpublished PhD thesis, University of York.

Foster, H. (1988), *Vision and Visuality.* (Seattle: New Press).

General Secretariat of the National Statistical Service of Greece (2007), 'Foreigners Arriving in Greece Classified by Means of Transport, Place of Entrance and Number of Cruise Passengers, January – December 2007', [Online: Hellenic Republic, Ministry of Economy and Finance]. Available at: www.statistics.gr [accessed: 12 September, 2008].

Hooper-Greenhill, E. (2000), *Museums and the Interpretation of Visual Culture.* (London: Routledge).

Howe, R.H. (1978), 'Max Weber's Elective Affinities: Sociology Within the Bounds of Pure Reason', *American Journal of Sociology* 8:2, 366–85.

Jay, M. (1996), 'Vision in Context: Reflections and Refractions', in M. Jay and T. Brennan (eds) *Vision in Context: Historical and Contemporary Perspectives on Sight*, 3–14. (London: Routledge).

Kaldis, W.P. and Lagoe, R.J. (1979), 'Background for Conflict: Greece, Turkey and the Aegean Islands, 1912–1914', *Journal of Modern History* 51:2, 1119–46.

Karousos, C. (1973), *Rhodos, History, Monuments, Art* (trans. Helen Dalambira). (Athens: 'Esperos' Editions).

Kinross, Lord (1979), *The Ottoman Centuries: The Rise and Fall of the Turkish Empire.* (London: Harper Perennial).

Kirshenblatt-Gimblett, B. (1998), *Destination Culture, Tourism, Museums and Heritage.* (Berkeley: University of California Press).

Kollias, E. (1998), *The Medieval City of Rhodes and the Palace of the Grand Master*. (Athens: Ministry of Culture Archaeological Receipts Fund).

Levine, M.A., Britt, K.M. and Delle, J.A. (2005), 'Heritage Tourism and Community Outreach: Public Archaeology at the Thaddeus Stevens and Lydia Hamilton Smith Site in Lancaster, Pennsylvania', *International Journal of Heritage Studies* 11:5, 399–414.

Lowenthal, D. (1998), *The Heritage Crusade and the Spoils of History*. (Cambridge: Cambridge University Press).

MacCannell, D. (1999), *The Tourist: A New Theory of the Leisure Class.* (Berkeley and Los Angeles: University of California Press).

MacDonald, F. (2002), 'The Scottish Highlands as Spectacle', in S. Coleman and M. Crang, *Tourism, Between Place and Performance*, 54–72. (Oxford: Berghahn Books).

Manis, T. (2008), 'Rhodes: An Architectural Traffic Jam', *Turkish Daily News* [Online, 12 July]. Available at: http://www.turkishdailynews.com.tr/article.php?enewsid=109675 [accessed 1 October 2008].

Marwick, K. (2001), 'Postcards from Malta: Image, Consumption, Context', *Annals of Tourism Research* 28:2, 417–38.

McLean, F. (2006), 'Introduction: Heritage and Identity', *International Journal of Heritage Studies* 12:1, 3–7.

Meethan, K. (2001), *Tourism in Global Society: Place, Culture, Consumption.* (Basingstoke: Palgrave).

Mellinger, W.M. (1994), 'Towards a Critical Analysis of Tourism Representations', *Annals of Tourism Research* 21:4, 756–79.

Merriman, N. (ed.) (2004), *Public Archaeology.* (London: Routledge).

Metz, C. (1982), *The Imaginary Signifier: Psychoanalysis and the Cinema*, trans. Celia Britton. (Bloomington: Indiana University Press).

Moser, S. (2001), 'Archaeological Representation, The Visual Conventions for Constructing Knowledge About the Past', in I. Hodder (ed.) *Archaeological Theory Today*, 262–83. (Cambridge: Polity Press).

Norwich, J.J. (1998), *A Short History of Byzantium.* (Harmondsworth: Penguin).

Osborne, P. (2000), *Travelling Light: Photography, Travel and Visual Culture.* (Manchester: Manchester University Press).

Palmer, C. (1999), Tourism and the Symbols of Identity, *Tourism Management* 20, 313–21.

Palmer, C. (2005), 'An Ethnography of Englishness: Experiencing Identity through Tourism', *Annals of Tourism Research* 32:1, 7–27.

Pretes, M. (2003), 'Tourism and Nationalism', *Annals of Tourism Research* 30:1, 125–42.

Rose, G. (1993), *Feminism and Geography: The Limits of Geographical Knowledge.* (Cambridge: Polity Press).

Schurman, J.G. (1916), *The Balkan Wars: 1912–1913.* (Princeton, NJ: Princeton University Press).

Selwyn, T. (ed.) (2006), *The Tourist Image: Myths and Myth Making in Tourism.* (London: Wiley).

Smith, L. (2006), *Uses of Heritage.* (London: Routledge).

Sorensen, L.W. and Pentz, P. (1992), *Lindos IV, 2. The Post-Mycenaean Periods until Roman Times and the Medieval Period.* (Copenhagen: The National Museum of Denmark).

Thomas, J. (2001), 'Archaeologies of Place and Landscape', in I. Hodder (ed.), *Archaeological Theory Today*, 165–86. (Cambridge: Polity Press).

Thurlow, C., Jaworski, A. and Ylänne-McEwen, V. (2005), 'Half-Hearted Tokens of Transparent Love? "Ethic" Postcards and the Visual Mediation of Host-Tourist Communication', *Tourism Culture and Communication* 5, 1–12.

Torr, C. (1885), *Rhodes in Ancient Times.* (Cambridge: Cambridge University Press).

Urry, J. (1990), *The Tourist Gaze.* (London: Sage Publications).

Vergo, P. (1989), 'The Reticent Object', in P. Vergo (ed.) *The New Museology.* (London: Reaktion Books).
Wang, N. (1999), 'Rethinking Authenticity in Tourism Experience', *Annals of Tourism Research* 26:2, 348–70.
Waterton, E. (2009), 'Sights of Sites: Picturing Heritage, Power and Exclusion', *Journal of Heritage Tourism* 4:1, 37–56.
Yüksel, A. and Akgül, O. (2007), 'Postcards as Affective Image Makers: An Idle Agent in Destination Marketing', *Tourism Management* 28, 714–25.

Index